Power Electronic Converter Harmonics

IEEE PRESS
445 Hoes Lane, P.O. Box 1331
Piscataway, NJ 08855-1331

Power Electronic Converter Harmonics

MULTIPULSE METHODS FOR CLEAN POWER

Derek A. Paice

IEEE Industry Applications Society, *Sponsor*

IEEE PRESS

The Institute of Electrical and Electronics Engineers, Inc., New York

This book may be purchased at a discount from the publisher when ordered in bulk quantities. For more information, contact:

IEEE PRESS Marketing
Attn: Special Sales
P.O. Box 1331
445 Hoes Lane
Piscataway, NJ 08855-1331
Fax: (732) 981-9334

Printed in the United States of America

10 9 8 7 6 5 4 3 2 1

ISBN 0-7803-5394-3
IEEE Order Number: PP5604

The Library of Congress has catalogued the hard cover edition of this title as follows:

Paice, D. A.
Power electronic converter harmonics : multipulse methods / Derek A. Paice.
p. cm.
Includes bibliographical references and index.
ISBN 0-7803-1137-X
1. Electric power supplies to apparatus. 2. Electric power system stability. 3. Harmonics (Electric waves) 4. Phase distortion (Electronics) I. Title.
TK7868.P6P35 1995
621.317—dc20 95-33342
CIP

Contents

Preface

Applications for solid-state equipment continue to increase, especially because of their ability to help conserve energy and provide better control of traditional and new processes. By the year 2000 it is estimated that fully 60% of all electrical power will be processed in some way by solid-state methods. The inherent nonlinear nature of the solid-state equipment load places harmonic current demands and extraneous losses upon the electrical power system. Means to control these demands are essential. In the United States, a major work that recommends limits on harmonic voltage and current was made available by the IEEE in May 1993. Prepared by experts in the Industry Applications Group, it is a valuable reference document and is likely to be considered by other standardization committees around the world. The document, IEEE Std 519-1992, is titled *IEEE Recommended Practices and Requirements for Harmonic Control in Electric Power Systems* (ISBN 1-55937-239-7). It can be obtained through the IEEE Headquarters, 345 East 47th Street, New York, New York 10017-2394 U.S.A. Within the industry it is often referred to simply as the IEEE Harmonics Spec.

By following the recommendations of IEEE Std 519-1992, the electric power system is expected to be of sufficient purity to be a useful source of power for all users of electricity. To meet the specifications requires appropriate integration of the Power Electronics equipment and the power source. It also requires design and measurement techniques that make it easy to determine when the recommendations are being met. A new design criterion is used in this book in that equipment designs are evaluated under two sets of conditions. The first is simply a classical balanced sinusoidal power source which I here define as "source power type 1" (SP1). This type of power is useful for analysis of basic equipment characteristics. The other type of power, defined as "source power type 2" (SP2), facilitates calculations under practical design and operating conditions. Power type SP2 incorporates 1% negative sequence voltage to simulate practical conditions of line voltage unbalance. Also a preexisting 2.5% 5th harmonic of voltage is included. This is representative of harmonic voltage that could be caused by a large 6-pulse rectifier. Calculations with power type SP2 are especially important when multiple parallel circuits are used to raise the effective pulse number of three-phase converters.

In following the IEEE recommendations three basic system approaches are considered in this book; specifically,

1. The installed power electronic load is restricted to a small amount of total load (generally not an acceptable solution).
2. The total power electronic load is filtered. This can be by passive or active filtering components or by phase staggering of various loads to obtain harmonic cancellation.
3. The individual power electronic loads are designed to curtail harmonic currents.

The great majority of power electronic equipment operates from an ac source but with an intermediate dc link. Thus, a significant opportunity exists to facilitate power electronics application by using ac to dc rectifiers that produce low harmonic current in the ac source. Multipulse converters can be applied to achieve this and these so-called "clean" power converters are of major interest in higher power ratings. Multipulse converters have been addressed in conventional treatises aimed at establishing basic performance features. However, these traditional methods of analysis have used closed form analyses for which certain simplifications are desirable to make the results easily interpreted. These simplifications are quite proper for establishing basic operation but may be the source of very significant errors when practical harmonic current measurements are made.

For example, in three-phase systems line voltage unbalance can produce significant amounts of 3rd harmonic current. Also preexisting harmonic voltages can cause an exaggerated amount of harmonic current. Further, the classical analyses which assume an ideal (ripple free) dc link current rarely apply. For the ubiquitous variable-frequency drive, ratings less than 500 HP are the norm, and harmonic current calculations in these cases must allow for the effects of finite dc link inductance and ripple in the dc link current.

All these factors can be accommodated by appropriate digital circuit analysis and resulting performance graphs that encompass practical variations of important parameters. By these means the extent to which power type SP2 modifies results is readily seen. Complete closed-form solutions could only be developed with great difficulty, but these would then need graphical interpretation for design use. The digital simulation approach is a simpler quicker method leading to the same results.

I believe that one of the greatest contributions anyone can make in engineering is to reduce a difficult subject to its simple essence. My early expectations were that computer simulations would detract from the kind of analogous thinking necessary to truly comprehend physical phenomena. In fact I have found quite the reverse to be true. Digital/graphical simulation is a tool that can simplify the understanding of many complex problems.

For example, investigation of a 12-pulse rectifier comprising two parallel rectifier bridges shows that it is highly sensitive to the phase angle of existing 5th harmonic voltages in the supply line. This is because the average dc output of each bridge is modified differently. Thus, the bridges become unbalanced. This simple result is easily observed in a digital simulation but would be more difficult to "find" using closed-form methods.

Several digital simulation software packages are commercially available. I found ECA-2 "Electronic Circuit Analysis" to be very effective and simple to apply for the power electronic circuits analyzed in this book. For those who wish to use ECA-2 for the simulation of power electronic circuits, arrangements for a special book-purchaser discount have been made with the supplier, Tatum Labs of Ann Arbor, Michigan. Free, limited-range demos are also available. To order, call 313/663-8810 and refer to: "Electronic Circuit Analysis" ECA-2, part no. E2PM-DPBK, as described in this book.

Because multipulse methods provide simple means to reduce harmonic currents at source they are covered extensively in this book. These circuits particularly apply to 480-V three-phase variable-frequency motor controllers. Phase-shifting transformers are used both with double-wound and autoconnections. A number of previously unpublished transformer arrangements and converter connections are presented. Analysis of these methods leads to design criteria that should be of help to those involved with design and application of three-phase power electronic equipment. Practical arrangements are presented that illustrate application of some of the new methods.

Ratings of single-phase converters are generally much lower than those of three-phase equipment; therefore, they have not been treated as individual "power" electronics in this book. However, in total, the single phase loads may represent a significant source of harmonic currents, especially the 3rd harmonic. This 3rd harmonic current is of particular concern in the neutral conductors of branch circuits and affects the transformer design. The problem is reviewed, and some of the solutions are briefly outlined.

Many of the arguments in favor of "cleaner" power electronics are of a qualitative nature. They have been stimulated in some cases by difficult operational problems caused by highly nonlinear loads. These loads have generated harmonic currents and voltages which caused interference with other equipment or have developed overheating due to harmonic losses. IEEE Std 519-1992 discusses some of these effects in a quantitative manner, but a qualitative discussion is also in order.

The IEEE specifications would always be met if it were feasible and economic to deploy only power electronic equipments that draw relatively clean power; however, this is not essential. Exceptions could be when the local conditions are known to tolerate high harmonics or the amount of electronics load is such that specifications can be met with the simplest of equipment. For example, a highly nonlinear load of 1 kVA drawn from a transformer rated for 100 kVA is unlikely to cause system problems, but if there are 60 such loads installed, intolerable system interactions or harmonic losses could occur.

The IEEE guidelines give voltage and current harmonic limits to address these issues. In the case of current harmonics, distortion is expressed relative to the total fundamental load current. This is important so as to not unduly penalize small amounts of harmonic load. As regards voltage distortion a limit of 5% is known from experience to give good system performance. However, it is not a number above which a sudden transition occurs. If the IEEE voltage distortion limits of 5% are sometimes exceeded, it is not likely that suddenly all other equipments will fail to function, but rather that the incidence of erratic behavior is likely to increase.

For a quantitative evaluation I know of no data relating the number of interference problems to the level of harmonic distortion; yet undoubtedly interference can be a major concern. Additional power losses in components such as cables, transformers, and motors are more readily analyzed and are discussed in IEEE Std 519-1992.

One argument for specifying cleaner power electronics is that of paving the way for future deployment of even more electronics without adding more system capacity. With some of the methods presented in this book, it is feasible for the total system load to be power electronics without exceeding harmonic specifications and without requiring any derating of the power system.

Finally, it should be noted that reduction of harmonic currents is incorporated at a cost. Because of the qualitative issues involved, it is difficult to make total system trade-offs for installation and operating costs. However, the relative values of these two factors can be determined for different harmonic mitigation techniques. This provides a useful format for comparison.

Acknowledgments

For the past 35 years, my life has been filled with electrical engineering. This book gives me a chance to share some of the knowledge I have acquired along the way. I hope it will be of help to others.

We grow with assistance from colleagues and friends, and I have enjoyed inputs and help from so many: at the aircraft development groups in the United Kingdom where I worked for 6 years, during my 20 years at the Westinghouse Research Center in Pittsburgh, and finally at the Westinghouse plant in Oldsmar, Florida, where I spent the last 9 years as engineering manager with a team involved in variable-frequency controllers. It was in this final assignment that the need and concepts for clean power took hold.

I am grateful for the support provided by Westinghouse Electric Corporation, the company that employed me for 29 years, and I am indebted to them for supplying practical data. Steve Marcum, Bill Vasiladiotis, and Ken Mattern helped in this regard. My colleague Valerie Tur reviewed parts of the book and helped me focus on issues of special interest to consulting engineers.

My associate Allan Ludbrook (of Ludbrook & Associates) contributed helpful comments, especially in regard to passive filter design.

Field test results were made available by Sam Terry of Norberg Engineering and William Cassity of Control Manufacturing Company, Inc. Their results support some of the calculations made in this book and lend credence to the work.

I especially want to acknowledge the support of two experts who have made numerous contributions in the field of power electronics and have been close colleagues for many years. Bob Spreadbury, a Westinghouse advisory engineer, freely contributed to design discussions and applied his many talents to produce final working designs. John Rosa, a Westinghouse engineering consultant, gave me encouragement, reviewed the complete manuscript, and offered valuable suggestions, most of which have been incorporated.

Finally, my dear wife Joan provided inspiration and support that made writing this book a pleasure. I dedicate it to her and my family, which includes Raymond, Anthony, and Annette.

Derek A. Paice
Palm Harbor, Florida

Nomenclature

OVERVIEW OF DEFINITIONS

The symbols in this book are defined as they are used and are repeated within some of the analyses to make it easier for the reader to follow. Word definitions follow those given in IEEE Std 519-1992, with clarification where it is thought to be helpful. Many of the letter definitions contained in IEEE 519 have also been used. In this book, lowercase symbols are used for instantaneous values. For example, i and v would refer to instantaneous values of current and voltage, respectively. The rms values for these quantities would use uppercase, that is I and V.

SUBSCRIPTS

Subscripts are used to identify the symbols to the extent it is possible. For example, I_L refers to the rms value of the line current. Subscripts are also used to define voltage potential. For example, $v_{1\text{-}2}$ is the instantaneous voltage difference between points 1 and 2. Positive voltage is defined when point 1 is positive with respect to point 2. Voltage $V_{1\text{-}2}$ would be the root-mean-square (rms) voltage acting from 1 through the external circuit to 2.

d = direct current or voltage

h = order of harmonic

L = line or line-to-line value

N = neutral point

sc = short circuit

o = open circuit, for example, V_{do}

in = input, for example, i_{in}

pk = crest value, for example, V_{pk}

pu = per unit quantity

fl = full load

LETTER SYMBOLS

Voltages

v = instantaneous voltage

V = rms voltage

V_P = rms value of phase voltage

V_L = rms value of line-to-line voltage

V_d = average value of dc voltage

V_{do} = average value of direct voltage at no load

$V_{L\text{-}N}$ = rms value of line-to-neutral voltage

V_H = total rms value of line-to-neutral harmonic voltage

V_h = rms value of voltage with harmonic number h

Currents

i = instantaneous current

I = rms current

I_L = maximum demand fundamental load current

I_l = rms load current

I_d = average value of dc current

I_{sc} = short-circuit amperes

Others

P = power

P_d = dc power

P_{fl} = power at full load

Φ = angle, in degrees or radians as appropriate

ω = angular frequency, in radians per second

f = frequency, in hertz

X_T = percent reactance, namely, $100 \times I_{fl}/I_{sc}$

$P_{do} = V_{do} I_d$

h = harmonic number

N = virtual and actual neutral

AT = ampere-turns

WORD DEFINITIONS

As per IEEE Std 519-1992 (see Section 1.5).

MULTIPULSE CIRCUIT DESCRIPTIONS

MC-101. Twelve-pulse with conventional delta-delta/wye double-wound transformer.

MC-102. Twelve-pulse with two identical auto-connected ±15° phase-shifting transformers and two identical interphase transformers.

MC-102A. A variation of MC-102 in which a single interphase transformer is used in conjunction with a zero sequence blocking transformer.

MC-103. Twelve-pulse with hybrid delta/wye transformer.

MC-104. Eighteen-pulse with differential fork step-down transformer.

MC-105. Twelve-pulse with differential fork step-down transformer.

Chapter 1

Harmonic Requirements

1.1 INTRODUCTION

Placing limits upon the effects that nonlinear loads may produce on users of electric power requires definition of system and equipment parameters. The IEEE Std 519–1992 document provides many of those definitions that are reproduced at the end of this chapter. They offer a standardized terminology that facilitates discussion of system harmonic issues. The basic requirements of voltage distortion and current distortion are guides for many users. When followed they eliminate most of the power system concerns relating to application of solid state equipment.

Telephone interference factor (TIF) is still under review, but the harmonics of voltage and current are critical parameters. By addressing these and conforming to IEEE Std 519–1992, some control of telephone interference is automatically provided.

The original IEEE Std 519 specification, issued in 1981, focused almost entirely on the matter of system voltage distortion, which is heavily dependent upon system characteristics. To determine voltage distortion, potential equipment suppliers often had to perform detailed system studies. Unwanted effects could be remedied by system as well as equipment changes; however, it was often unclear who should change what.

In the revised IEEE Std 519–1992 document the harmonic currents drawn by a users' equipment are also defined. This is something that manufacturers can address in equipment design. They are doing so, and it is hoped this book will provide additional help for designers and users. There is still a system factor involved because tolerable harmonic currents are defined relative to the total system load. This is as it should be; however, the system definition can be less detailed for this and performance expectations are more readily determined.

1.2 VOLTAGE DISTORTION

Voltage distortion defines the relationship between the total harmonic voltage and the total fundamental voltage. Thus, if the fundamental ac line to neutral voltage is $V_{L\text{-}N}$ and the total line to neutral harmonic voltage is V_H, then

$$\text{total harmonic voltage distortion} = \frac{V_H}{V_{L-N}} \tag{1.1}$$

where

$$V_H = \sqrt{\sum_{h=2}^{h=25} V_h^2} \tag{1.2}$$

An upper summation limit of $h = 25$ is chosen for calculation purposes. It gives good practical results. Recommended voltage distortion limits are summarized in Table 1-1.

TABLE 1-1.

CLASSIFICATION AND VOLTAGE DISTORTION LIMITS FOR INDIVIDUAL USERS (LOW-VOLTAGE SYSTEMS)

Class of System	Total Harmonic Distortion	Notch Area Volt—μsec*	Notch Depth
†Special applications	3%	16,400	10%
General system	5%	22,800	20%
Dedicated system	10%	36,500	50%

*Multiply this value by $V/480$ for other than 480 V systems.
†Special applications include hospitals and airports.

1.2.1 Line Notching Calculations and Limits

Notching refers to the effects that commutation has on the ac line voltage. It is most easily demonstrated with respect to a 6-pulse three-phase converter bridge with dc filter inductor as shown in Figure 1-1.

Line notching results when two semiconductors of the same polarity are simultaneously contributing to the load current I_d. This occurs, for example, when device S_2 starts to conduct current and supply the current previously supplied through device S_1. Changeover of current conduction from one device to the other takes time (commutation time), and during that interval the voltage difference between line A and line B is zero because the devices S_1 and S_2 ideally have no voltage drop. Figure 9-1 shows currents during commutation in detail.

With diode operation (gating at 0° in Figure 1-1), the current changes naturally from one device to the other, and the notch has one fast rising side. With

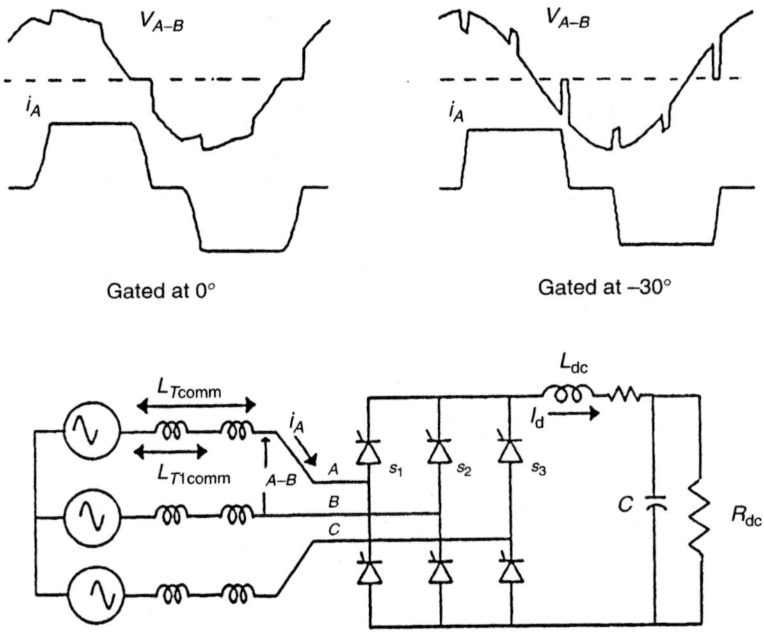

Figure 1-1 Illustrating line voltage notching effects.

SCR phase back, such as $-30°$ in Figure 1-1, the notch will have two fast-changing sides. This gives a greater likelihood of high-frequency interference.

When the converter is operated from the line without phase shift, as in Figure 1-1, the line voltage shows one large notch and two smaller notches. The IEEE Std 519–1992 specification defines the volt-seconds area relative to the larger notch.

During commutation the inductance $L_{T\text{comm}}$ in one line has a change in current from I_d to zero. In the other phase, a similar change of current I_d occurs except that this current goes from zero to I_d. To change current in a linear inductor requires an expenditure of volt seconds equal to the product of inductance and current. These volt-seconds are subtracted from the source line voltage. At the converter terminals in Figure 1-1 the notch area is ideally given by

$$\text{notch volt-seconds} = 2\,L_{T\text{comm}} I_d \tag{1.3}$$

At a point closer to the source, where for example the inductance back to the source is only $L_{T1\text{comm}}$, the corresponding notch volt-seconds are reduced by the ratio $L_{T1\text{comm}}/L_{T\text{comm}}$.

1.2.2 Notching with Delta/Wye Transformers

If the converter is operated from a phase-shifting transformer, the pattern of notching changes. In the case of a delta/wye transformer with 30° phase shift, there will now be two large notches. Waveforms for this example are shown in Figure 1-2.

The apparently simple notch calculation in equation (1.3) has to be modified for delta/wye systems. Specifically, the 2 multiplier becomes $\sqrt{3}$, as shown in Figure 1-3, page 6.

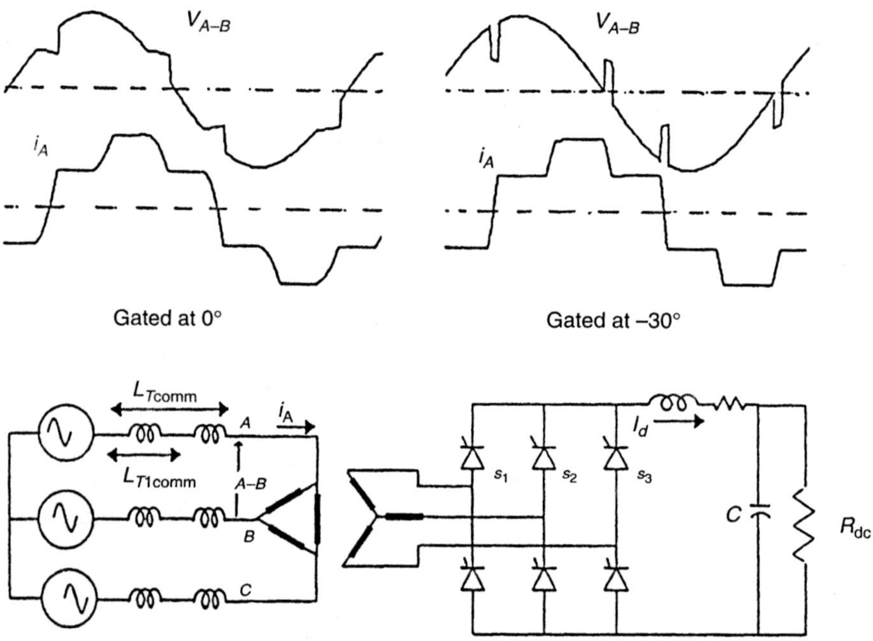

Figure 1-2 Showing the different notch pattern when converter is fed from delta/wye transformer

1.2.3 Effects of dc Filter Inductance on Notching

Practical 6-pulse converters often have a significant amount of ripple in the dc current, and at the instant of commutation, the current is usually less than the average value of I_d assumed in equation (1.3). When the dc inductance is negligible, the voltage waveshapes and notch patterns are greatly changed. For example, in diode bridge simulations for this case the major notch area was 28% less than that predicted by equation (1.3) when the ac line reactance was 6% and 38% less for 3% reactance. For 6-pulse phase-controlled converters, dc inductance is essential and notch patterns are more predictable. The notching limits specified in IEEE Std 519–1992 are reproduced in Table 1-1.

1.2.4 System Notches Caused by Multiple Converter Loads

In general the system load will include a mix of converter types and ratings. In this event the commutation of one unit affects commutation of another, and it is not possible to determine the total notch area by summing the effects of individual units. For analysis, a full-scale computer simulation is recommended to calculate the line voltage waveshapes. The author has his own favorite methods; however, various software packages are available. Practical oscillograms of system ac line voltages can be similarly analyzed to determine notch effects.

1.2.5 Notching in Multipulse Circuits

Multipulse circuits are introduced in Chapter 3 as a means for filtering power converters to produce more nearly sinusoidal currents. With smoother currents, the concept of notch volt-seconds becomes less viable. In the limit, when there are so many pulses that the line current is completely sinusoidal, there is still a voltage drop in the source reactance. This volt-seconds loss in each half cycle can be much greater than the IEEE Std 519–1992 limits without causing operating problems.

Multipulse arrangements cause less individual notch area than do 6-pulse circuits. For example, in a 12-pulse circuit formed from two 6-pulse circuits, the load current being commutated is reduced to one-half. Intuitively we would expect the notch area to be reduced by a factor of about 2; Figure 1-3 shows one example. In multipulse systems with reduced device conduction, the commutating reactance in the phase-shifting transformer secondary is greatly magnified when it is referred to the source voltage. The result is a nearly sinusoidal source current and negligible notch effects. Exact analysis of notches is of limited use. In specific cases, a simulation provides accurate results.

Another method for estimating notching considers the net effective change in current when steps of current occur. For example, consider the 6-pulse and 12-pulse currents i_1 and i_2 in Figure 1-3. At commutation the volt-secs absorbed will be the appropriate L_{comm} times the current change in each line. Specifically, for balanced ac line inductance:

$$\text{total notch volt-secs} = L_{\text{comm}}\,(\text{step in } i_1 - \text{step in } i_2)$$

This technique is illustrated in Figure 1-3. The concept can be applied to any converter circuit.

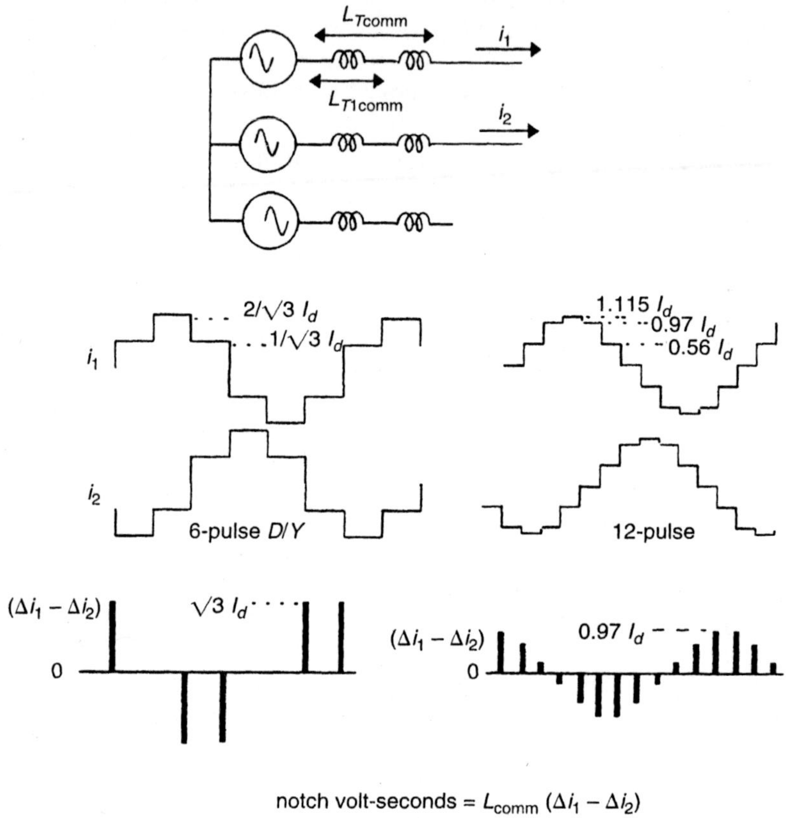

Figure 1-3 Calculating notch volt-secs from current steps.

1.3 CURRENT DISTORTION

In general, current distortion defines the relationship between the total harmonic current and the fundamental current in much the same way as voltage distortion. However, there are some application differences which need to be recognized. These include

- Current harmonic limits depend upon the system short-circuit current capability at the point of interest.
- Current harmonic percentages apply to individual harmonic currents. They are expressed relative to the total system fundamental load current

for worst case normal operating conditions lasting more than one hour. (They are not expressed relative to the fundamental current load of the nonlinear equipment.) The worst case operating conditions are expressed relative to the average current of maximum demand, preferably for the preceding 12 months.

- Total demand distortion TDD is the total harmonic current distortion given by

$$TDD = \frac{I_H}{I_L} \tag{1.4}$$

where I_L is the maximum demand load current (fundamental frequency component) at the PCC derived from a 15-minute or 30-minute billing demand kVA. And I_H is given by

$$I_H = \sqrt{\sum_{h=2}^{h=25} I_h^2} \tag{1.5}$$

The upper summation limit of $h = 25$ is chosen for calculation purposes. It gives good practical results.

The system harmonic current limits recommended in IEEE Std 519–1992 are shown in Table 1-2 for 6-pulse systems. For higher-pulse numbers, larger characteristic harmonics are allowed in the ratio (pulse number/6),$^{0.5}$ provided that noncharacteristic harmonics are less than 25% of the limits specified in the table.

TABLE 1-2. *Table 10.3 in IEEE Std 519–1992. Reprinted with permission.*

Maximum Harmonic Current Distortion in Percent of I_L						
	Individual Harmonic Order (Odd Harmonics)					
I_{sc}/I_L	<11	$11\le h<17$	$17\le h<23$	$23\le h<35$	$35\le h$	TDD
<20*	4.0	2.0	1.5	0.6	0.3	5.0
20<50	7.0	3.5	2.5	1.0	0.5	8.0
50<100	10.0	4.5	4.0	1.5	0.7	12.0
100<1000	12.0	5.5	5.0	2.0	1.0	15.0
>1000	15.0	7.0	6.0	2.5	1.4	20.0

Even harmonics are limited to 25% of the odd harmonic limits above.

Current distortions that result in a dc offset, e.g., half-wave converters, are not allowed.

*All power generation equipment is limited to these values of current distortion, regardless of actual I_{sc}/I_L.

where

I_{sc} = maximum short-circuit current at PCC.
I_L = maximum demand load current (fundamental frequency component) at PCC.

1.3.1 Current Distortion and Transformers

The previous discussion regarding total demand distortion (TDD) and calculation of harmonic current distortion relates to the IEEE Std 519–1992 specification. Another specification, ANSI/IEEE Std C57.110-1986, relates to the effect that harmonic currents have on power transformers covered by ANSI/IEEE C57.12.01-1979 and to power transformers up to 50-MVA maximum nameplate rating covered by ANSI/IEEE Std C57.12.00-1987. This specification does not apply to rectifier or special transformers. In these specifications, a definition of "harmonic factor" is used for current distortion. It is used to determine the transformer rating when the harmonic factor exceeds 0.05 per unit. This harmonic factor relates to the ratio of the effective harmonic current to the fundamental current. Thus, for transformer rating, it may not be sufficient to determine the system total demand distortion. The issue of transformer derating is dealt with in Chapter 7, Section 7.7.

1.4 TELEPHONE INTERFERENCE

There is as yet no formal specification for the allowable telephone influence factor; however, two formulas for calculation are given in IEEE Std 519–1992. Each uses the 1960 curves for telephone interference weighting factor. This factor takes into account the response of telephone sets and the human ear. Also, each formula directly incorporates line harmonic currents up to 5000 Hz.

One useful form of the formulas, for gauging the possibility of telephone interference, is given by the root-sum-square (RSS) of the product of individual harmonic current I_h and telephone interference factor T_h, namely, $I \bullet T$. It is given by

$$I \bullet T = \sqrt{\Sigma_{h=1}^{H}(I_h T_h)^2} \tag{1.6}$$

where H corresponds to 5000 Hz.

Specific frequency values for T_h up to the 49th harmonic of a 60-Hz converter are given in Table 1-3. Practical results from standard waveform analyzers cover this range. A larger range for TIF, up to 5000 Hz, is given in IEEE Std 519–1992. Table 1-4 gives calculated results for $I \bullet T$ using the idealized current amplitude harmonics of $1/(kq \pm 1)$, where q is the pulse number and k is any positive integer. Higher pulse number converters are seen to reduce the possibility of telephone interference.

In practice, results are much reduced because of the filtering affects of equipment reactance. For 50 Hz converters, the idealized $I \bullet T$ factors are reduced by approximately 8 percent.

TABLE 1-3.

SINGLE-FREQUENCY TIF (T_h) VALUES FOR HARMONICS OF 60 HZ					
***h*#**	**TIF**	***h*#**	**TIF**	***h*#**	**TIF**
1	0.5	17	5,100	31	7,820
2	15	18	5,400	33	8,330
3	30	19	5,630	35	8,830
5	225	21	6,050	36	9,080
7	650	23	6,370	37	9,330
9	1,320	24	6,560	39	9,840
11	2,260	25	6,680	41	10,340
12	2,760	27	6,970	43	10,600
13	3,360	29	7,320	47	10,210
15	4,350	30	7,570	49	9,820

TABLE 1-4.

IDEALIZED *I* • *T* FACTORS FOR 60-HZ CONVERTERS (PER FUNDAMENTAL AMPERE) (UP TO 49TH HARMONIC)		
6-Pulse	**12-Pulse**	**18-Pulse**
997	705	594
954	686	552*

*From measured data, practical 18-pulse equipments produce an *I* • *T* that is only 18% to 31% of this value. For example, a 480-V, 125-A, 18-pulse converter causes an *I* • *T* of 12,000. Referred to a 12-kV bus, this reduces to *I* • *T* = 480.

1.5 DEFINITIONS OF TERMS

Selected definitions in this section are from IEEE Std 519–1992. They are reproduced with permission of the Institute of Electrical and Electronic Engineers.

Definitions given here are tailored specifically to the harmonics generated by static power converters at utility system frequencies. Additional useful guidelines will be found in IEEE Std 100–1992, IEEE Std 223–1966, IEEE Std 59–1962, ANSI Std C34.2 1968, and IEEE Std 444–1973.

commutation. The transfer of unidirectional current between thyristor (or diode) converter circuit elements that conduct in succession.

converter. A device that changes electrical energy from one form to another. A semiconductor converter is a converter that uses semiconductors as the active elements in the conversion process.

distortion factor (harmonic factor). The ratio of the root mean square of the harmonic content to the root mean square of the fundamental quantity, expressed as a percentage of the fundamental.

$$\mathrm{DF} = \sqrt{\frac{\text{sum of squares of amplitudes of all harmonics}}{\text{square of amplitude of fundamental}}} \bullet 100\%$$

filter. A generic term used to describe those types of equipment whose purpose is to reduce the harmonic current or voltage flowing in or being impressed upon specific parts of an electrical power system or both.

filter, damped. A filter generally consisting of combinations of capacitors, inductors, and resistors that have been selected in such a way as to present a low impedance over a broad range of frequencies. The filter usually has a relatively low Q (X/R).

filter, high-pass. A filter having a single transmission frequency extending from some cutoff frequency, not zero, up to infinite frequency.

filter, series. A type of filter that reduces harmonics by putting a high series impedance between the harmonic source and the system to be protected.

filter, shunt. A type of filter that reduces harmonics by providing a low impedance path to shunt the harmonics away from the system to be protected.

filter, tuned. A filter generally consisting of combinations of capacitors, inductors, and resistors that have been selected in such a way as to present a relatively minimum (maximum) impedance to one or more specific frequencies. For a shunt (series) filter the impedance is a minimum (maximum). Tuned filters generally have a high Q (X/R).

harmonic. A sinusoidal component of a periodic wave or quantity having a frequency that is an integral multiple of the fundamental frequency. Note, for example, a component, the frequency of which is twice the fundamental frequency, is called a second harmonic.

harmonic, characteristic. Those harmonics produced by semiconductor converter equipment in the course of normal operation. In a six-pulse converter the characteristic harmonics are the nontriple odd harmonics, for example, the 5th, 7th, 11th, and 13th.

harmonic, characteristic. (continued for a 6-pulse converter)

$h = kq \pm 1$

$k =$ any integer

$q =$ pulse number of converter

harmonic, noncharacteristic. Harmonics that are not produced by semiconductor converter equipment in the course of normal operation. These may be a result of beat frequencies, a demodulation of characteristic harmonics and the fundamental, or an imbalance in the ac power system, asymmetrical delay angle, or cycloconverter operation.

harmonic factor. The ratio of the RSS value of all the harmonics to the rms of the fundamental.

$$\text{harmonic factor (for voltage)} = \frac{\sqrt{E_3^2 + E_5^2 + E_7^2 \ldots}}{E_1}$$

$$\text{harmonic factor (for current)} = \frac{\sqrt{I_3^2 + I_5^2 + I_7^2 \ldots}}{I_1}$$

***I* • *T* product.** The inductive influence expressed in terms of the product of its rms magnitude (I), in amperes, times its telephone influence factor.

***kV* • *T* product.** Inductive influence expressed in terms of the product of its rms magnitude, in kilovolts, times its telephone influence factor.

line voltage notch. The dip in the supply voltage to a converter due to the momentary short circuit of the ac lines during a commutation interval. Alternatively, the momentary dip in supply voltage caused by the reactive drops in the supply circuit during the high rates of change in currents occurring in the ac lines during commutation.

nonlinear load. A load that draws a nonsinusoidal current wave when supplied by a sinusoidal voltage source.

notch depth. The average depth of the line voltage notch from the sine wave of voltage.

notch area. The area of the line voltage notch. It is the product of the notch depth, in volts, times the width of the notch in microseconds.

power factor, displacement. The displacement component of power factor; the ratio of the active power of the fundamental wave, in watts, to the apparent power of the fundamental wave, in volt-amperes (including the exciting current of the converter transformer).[1,2]

pulse number. The total number of successive nonsimultaneous commutations occurring within the converter circuit during each cycle when operated without phase control. It is also equal to the order of the principal harmonic in the direct voltage, that is, the number of pulses present in the dc output voltage in one cycle of the supply voltage.

short-circuit ratio. For a semiconductor converter, the ratio of the short-circuit capacity of the bus, in MVA, at the point of converter connection, to the rating of the converter, in megawatts.

telephone influence factor (TIF). For a voltage or current wave in an electric supply circuit, the ratio of the square root of the sum of the squares of the weighted rms values of all the sine wave components (including alternating current waves both fundamental and harmonic) to the rms value (unweighted) of the entire wave.

total demand distortion (TDD). The total RSS harmonic current distortion, as a percentage of the maximum demand load current (15- or 30-minute demand).

total harmonic distortion (THD). This term has come into common usage to define either voltage or current "distortion factor." See distortion factor.

1.6 OTHER HARMONIC SPECIFICATIONS

The work in this book focuses on power systems used in the United States. For this reason the primary specification addressed is IEEE Std 519–1992.

In the United Kingdom, specification G.5/3 is a harmonics specification in the form of an engineering recommendation from The Electricity Council Chief Engineers' Conference. It is titled "Limits for Harmonics in the United Kingdom Electricity Supply System." In addition to specifying various harmonic limits, it precludes the use of certain power levels of converter equipment in different power systems. Some of the power equipment designs in this book, which conform to IEEE Std 519–1992, will be very effective at addressing the concerns of the G.5/3 specification.

[1]This definition includes the effect of harmonic components of current and voltage (distortion power factor), the effect of phase displacement between current and voltage, and the exciting current of the transformer. Volt-amperes are the product of rms voltage and rms current.

[2]The power factor is determined at the line terminals of the converter.

The International Electrotechnical Commission (IEC) published a first edition of publication IEC 555-2 in 1982 to address the impact of electrical equipment and appliances used in the home. The specification is entitled "Disturbances in Supply Systems Caused by Household Appliances and Similar Electrical Equipment, Part 2, Harmonics." CENELEC approved IEC-555-2 as a European standard (EN 60555-2) in December 1991. It includes individual single-phase equipment up to 16 A and is important to those involved in European markets. At the time of this writing, the standard is under revision [24].

It is desirable to address the generation of power line harmonics by consumer and professional electronic equipments; however, most of the equipment is single phase. In this case, individual 120-V ratings are limited to 15 A. Individually, these types of equipment are not considered as "power" electronics in this book. However, when large quantities of single-phase equipment are connected to a power system, significant harmonic currents result. The 3rd harmonic of current is especially of concern and is important in determining the ampacity of the neutral conductor in branch circuits. Discussion on this is provided in Chapter 7, Section 7.8.

Military specifications such as MIL-STD-1399 (NAVY) SECTION 300A also define acceptable levels of harmonic current generation.

Chapter 2

Power Source Representation

2.1 INTRODUCTION

IEEE Std 519–1992 specifies limits on the amount of various harmonic currents that can be accepted in power systems. Thus it is desirable to calculate harmonic line currents generated by various types of converter equipment under practical conditions. Classical analyses give basic performance of various circuit topologies; however, these results may be inadequate for practical design variations in equipment and power system. By focusing on power system variations and their effects, the ability to meet specifications such as IEEE Std 519–1992 is greatly improved. Some of the results may already be understood, such as the effects of negative-sequence voltage in producing 3rd harmonic current. Other results, such as serious unbalance in multipulse parallel converters due to preexisting voltage harmonics, are not well known.

Two types of power source characteristics are defined in this book for use in equipment design evaluation, namely, source power type 1 (SP1) and source power type 2 (SP2). Evaluating performance in these environments gives insight into sensitive design features and facilitates design of practical production equipment that meets IEEE harmonic requirements. Different industries may want to expand the parameters to better suit their application and design needs. However, source power type SP2 is believed to be representative of many practical situations.

2.2 DEFINITION OF VOLTAGE VECTORS

Sometimes it is advantageous to show vectors with different frequencies on the same diagram. This procedure is not without pitfalls, and a brief discussion may be helpful.

A vector drawing usually shows the relative peak amplitude and phase relationship of voltages and currents of the same frequency, but with appropriate interpretation it can also display harmonic vectors. The fundamental reference vector is shown as a straight line parallel to the *X* axis. It rotates about the left end, in a counterclockwise direction for positive angles and clockwise for negative angles. The rate of rotation is expressed as $2\pi f$ radians per sec or ω. At any instant the projection of the vector on the *Y* axis gives the instantaneous value of the sine wave represented by that vector.

In a three-phase system with balanced symmetrical voltages the sequence is defined as positive when the vectors reach their peak amplitudes in the order *A, B, C.* If, with the same counterclockwise rotation, the vectors reach their peaks in the order *A, C, B,* as it does for the 5th harmonic in Figure 2–1, the sequence is defined as negative. An important property of these voltage sets is that if a positive-sequence set experiences a phase shift of $+\Phi$ in passing through a transformer, then a negative-sequence set will undergo a phase shift of $-\Phi$.

The relative phase of each vector is conveniently defined with respect to the reference phase, at angle $\omega t = 0$. All vectors can be expressed relative to this reference phase, including different frequency (harmonic) vectors. However, as the vectors rotate, they do so at different rates. For example, if the fundamental moves through an angle ωt (radians), then a 5th harmonic will move through an angle $5\omega t$.

Figure 2-1 shows two methods of representing different frequency voltage vectors. For mixed frequencies, the nondimensional time diagram showing voltage amplitude as a function of ωt is often easier to follow than the classical vector (phasor) drawing.

2.3 SOURCE POWER TYPE 1

Source power type 1 (SP1) is defined as a three-phase voltage source with balanced voltages, balanced inductive impedances, and no preexisting harmonic voltages. If the load is nonlinear, the resulting voltage distortion will depend only upon the harmonic currents generated and the source impedance. This type of power is idealized and really only useful for analysis of basic equipment features. It is most often used in classical closed-form analysis.

2.4 SOURCE POWER TYPE 2

Without any information to the contrary, source power type 2 (SP2) is recommended to be used as a minimum for practical equipment design. It is defined as a three-phase voltage source with balanced inductive impedances but with 1%

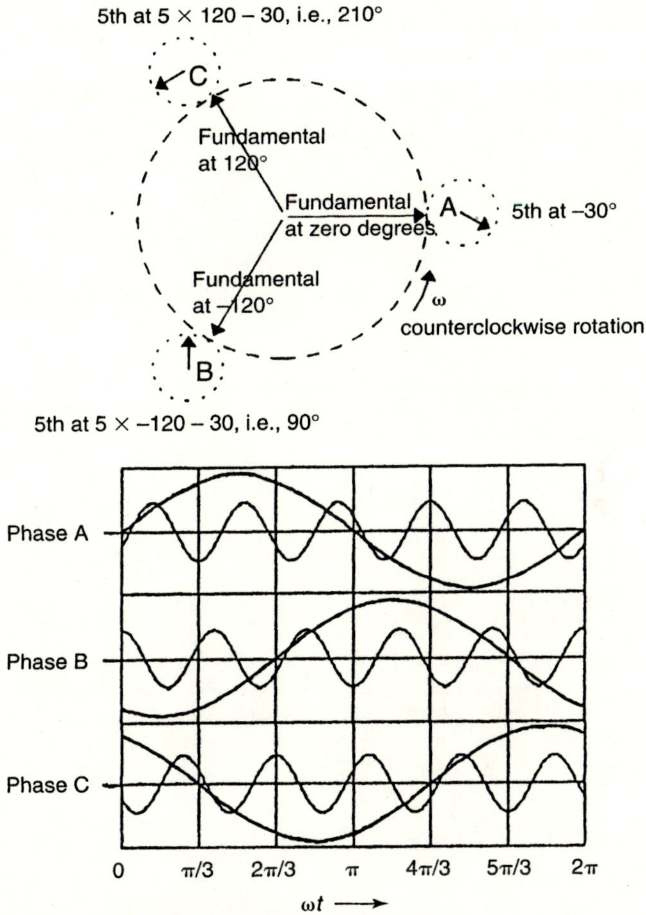

Figure 2-1 Relationships of fundamentals with 5th harmonics at angle 30°.

voltage unbalance (negative sequence). Also, a preexisting 5th harmonic voltage of 2.5% is included.

It was originally thought to include in this type of power a full spectrum of frequencies characteristic of 6-pulse converter systems, for example, $6k \pm 1$. However, it was found that consideration of only the 5th harmonic was sufficient to highlight important sensitivities. As an example, the 5th harmonic is shown to have significant impact in the performance of multipulse systems. Choosing topologies insensitive to this harmonic will result in designs insensitive to many other harmonics.

2.5 SUMMARY OF POWER TYPES

Table 2-1 shows how the phase voltages are represented for type 1 and type 2 power. In both cases the source impedance is assumed to be inductive and balanced. Thus it can be characterized by an appropriate short-circuit current. The effects of unbalance in the source impedance values can be incorporated within a digital simulation; however, in most designs source impedance variations of ±5% have little effect on individual harmonic currents. IEEE Std 519–1992 indicates that most systems will have source impedance variations of less than ±5%.

TABLE 2-1.

REPRESENTATION OF PHASE VOLTAGES FOR POWER TYPES 1 AND 2*,†			
Voltage	**Phase *A***	**Phase *B***	**Phase *C***
Fundamental amplitude = $V1$	$\sin \omega t$	$\sin(\omega t - \frac{2\pi}{3})$	$\sin(\omega t + \frac{2\pi}{3})$
Negative-sequence amplitude = 0.01 $V1$	$\sin \omega t$	$\sin(\omega t + \frac{2\pi}{3})$	$\sin(\omega t - \frac{2\pi}{3})$
5th harmonic amplitude = 0.025 $V1$	$\sin(5\omega t + \phi)$	$\sin(5\omega t + \frac{2\pi}{3} + \phi)$	$\sin(5\omega t - \frac{2\pi}{3} + \phi)$

*For type 1 power, set negative-sequence and 5th harmonic to zero.

†The phase position ϕ is defined relative to the harmonic when the fundamental is in the reference position $(1 + j0)$.

2.6 SOME EFFECTS OF SOURCE POWER TYPE 2

Complete closed-form solutions are rarely feasible when practical system variations are to be included. Nevertheless, using appropriate assumptions, some simplified analyses are possible that highlight effects of some of the variables. For complete results, digital solutions are then obtained and the results graphed.

The basic "building unit" for many practical multipulse converters is a three-phase rectifier bridge with dc filter inductor. In this circuit the semiconductors pass current for nominally 120° periods. In turn, the bridge circuit is comprised of an even more basic arrangement using three semiconductors in a 3-pulse midpoint connection [1, 2]. These two basic circuits are shown in Figures 2-2 and 2-3.

The circuit of Figure 2-2 will be used to determine some of the effects of negative-sequence and preexisting harmonic voltages. These results also apply to the combination of two of these circuits which are effectively connected in series to form a three-phase bridge converter.

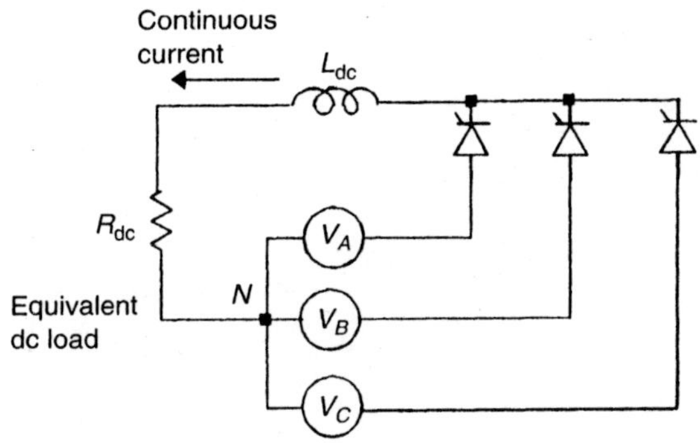

Figure 2-2 A 3-pulse midpoint connection for analysis.

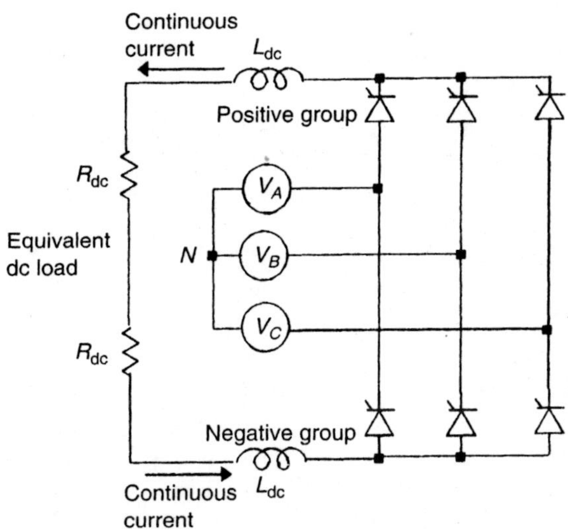

Figure 2-3 Two 3-pulse midpoint connections in series to form a three-phase bridge.

2.6.1 Effects of Negative-Sequence Voltages

The effect of negative-sequence voltage is not identical in each pulse of the dc output voltage. Therefore, results must be calculated over each conduction period to find the total effect on the converter dc output. The amount of negative-sequence voltage is small, only 1%, and its impact on the instant of commutation has been neglected.

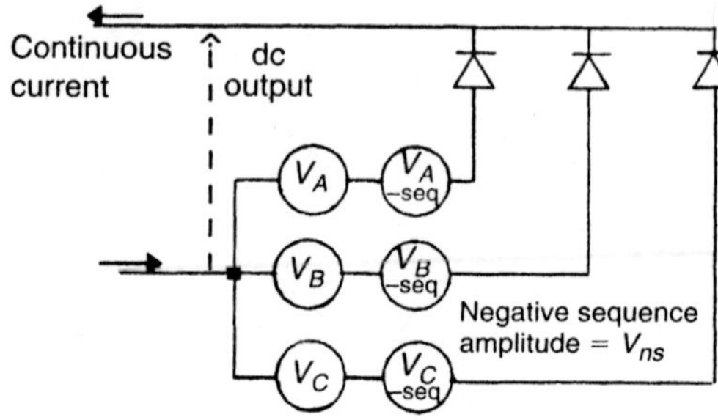

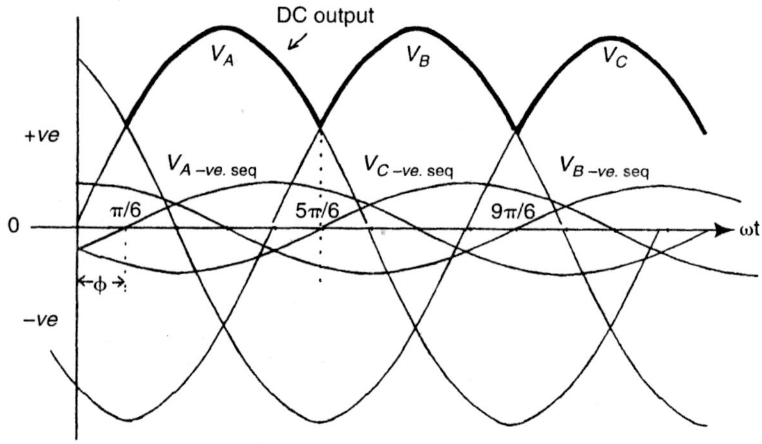

Figure 2-4 Showing positive and negative-sequence voltages for analysis. (Note: Negative-sequence voltages are greatly amplified.)

The dc voltage V_{dcns} produced by the negative-sequence voltage V_{ns} is the sum of three pulses over a period of 2π. Thus, from Figure 2-4,

$$\frac{2\pi}{V_{ns}} V_{dcns} = \int_{\pi/6}^{5\pi/6} sin(\omega t + \phi) + \int_{5\pi/6}^{9\pi/6} sin\left(\omega t + \frac{2\pi}{3} + \phi\right) + \int_{9\pi/6}^{13\pi/6} sin\left(\omega t - \frac{2\pi}{3} + \phi\right)$$

$$\frac{2\pi}{\sqrt{3V_{ns}}} V_{dcns} = sin\left(\frac{\pi}{2} + \phi\right) + sin\left(11\frac{\pi}{6} + \phi\right) + sin\left(7\frac{\pi}{6} + \phi\right)$$

$$= 0$$

These equations show that, for the stated conditions, the net effect on the 3-pulse converter dc output is zero. However, individual output pulses of the converter bridge are differently affected over each conduction period. Considering two 3-pulse converters in series to form a 6-pulse bridge as shown in Figure 2-3, the net dc output still averages to zero. However, the different average value of each pulse results in a 2nd harmonic of voltage in the dc output of a three-phase bridge connection. This, in turn, causes a 3rd harmonic in the input line current. A quantitative evaluation of this effect is left for the digital simulations.

In summary, the presence of negative-sequence voltage in the power source does not affect the dc output but causes 3rd harmonic in the converter line current.

2.6.2 Effects of Preexisting Harmonic Voltages

Figure 2-5 shows the reference phase A voltage applied to the basic 3-pulse circuit and includes a 5th harmonic voltage with phase lag of ϕ.

It can be observed from this figure that the 5th harmonic of voltage alternately adds to and subtracts from the dc output over the conducting period of $2\pi/3$. For simplicity the effects of commutation angle changes due to the harmonic voltage are neglected in this initial analysis.

In general, the power system is at least expected to contain harmonic voltages related to the characteristic harmonics ($6k \pm 1$) of 6-pulse rectifiers. Thus the general phase A voltage V_A can be represented by

$$V_A = V_1 \sin \omega t + \sum_{k=1}^{k=\infty} V_{(6k\pm1)} \sin[(6k \pm 1)\omega t + \phi_{(6k\pm1)}] \tag{2.1}$$

where k is any positive integer.

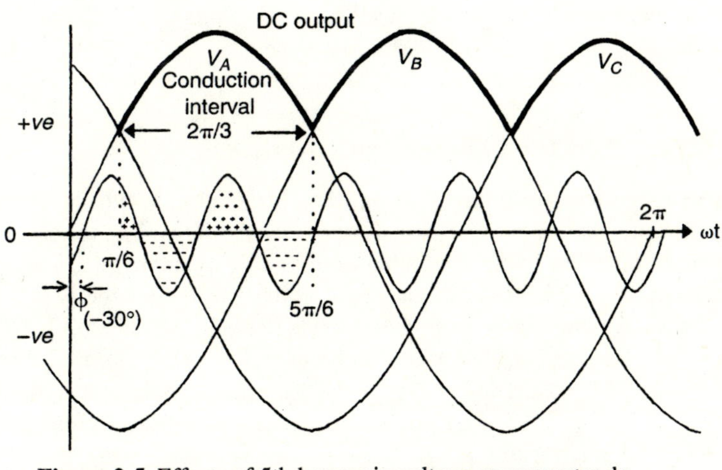

Figure 2-5 Effects of 5th harmonic voltage on converter dc output. (Note: Scale for 5th harmonic is greatly magnified.)

The essential features of this analysis can be presented by considering only the first two of these characteristic harmonics, namely, the 5th and 7th. These occur when $k = 1$, and in this case

$$V_A = V_1 \sin \omega t + V_5 \sin\left(5\omega t + \phi_5\right) + V_7 \sin\left(7\omega t + \phi_7\right) \tag{2.2}$$

The converter dc output is similarly affected by harmonic voltage in each $2\pi/3$ conduction period; thus,

$$V_{dc} = \frac{3}{2\pi} \int_{\pi/6}^{5\pi/6} \left[V_1 \sin \omega t + V_5 \sin\left(5\omega t + \phi_5\right) + V_7 \sin\left(7\omega t + \phi_7\right)\right] dwt \tag{2.3}$$

Completing this integration leads to

$$V_{dc} = \frac{3\sqrt{3}}{2\pi} V_1 \left(1 - \frac{1}{5}\frac{V_5}{V_1}\cos\phi_5 - \frac{1}{7}\frac{V_7}{V_1}\cos\phi_7\right) \tag{2.4}$$

When two 3-pulse units are series connected to form a three-phase bridge the amplitude of V_{dc} in (2.4) is doubled.

These results show that preexisting 5th and 7th harmonics of voltage in the power source cause the dc output of a 6-pulse converter bridge to be affected by an amount dependent upon the harmonic phase angle.

If the preexisting 5th harmonic voltage is 2.5%, then the dc output of a single 6-pulse converter bridge can be changed by as much as ±0.5% depending upon the harmonic phase angle. If both 5th and 7th are present in the amount of 2.5%, the output of a single bridge may change by ±0.857%. Thus, for two phase-shifted converter bridges in parallel, the dc voltage difference can be as much as ±1% with 2.5% of 5th harmonic present and ±1.71% when both 5th and 7th are present in the amount of 2.5%.

2.6.3 Effects on Multipulse Converters

Multipulse converters use phase shifting transformers to provide cancellation of certain harmonic currents. The transformers also affect the phase of power source line voltage harmonics, but in an unfavorable manner.

For example, assume the 5th harmonic voltage in the source is at angle 0°. For a 12-pulse system with a second converter fed at +30°, the 5th harmonic in the supply for the second converter will be at $-5 \times 30 - 30$, that is, $-180°$. (Advance phase A fundamental in Figure 2-1 by $\pi/6$ to see this.) A 2.5% 5th harmonic voltage will cause the dc output of the first converter to decrease 0.5% and the second converter to increase 0.5%.

If, for example, two 620 V dc converters are paralleled, there will be a dc voltage unbalance of 6.2 V. If this voltage is not corrected by regulating means, it will cause significant current unbalance. In turn this reintroduces 6-pulse harmonics to the line currents. These effects are clearly shown in Figure 2-6.

One might anticipate from these results that design of interphase transformers is quite complicated, and so it is. Additionally, it appears that the performance of typical 12-pulse systems can easily be affected by changes in the

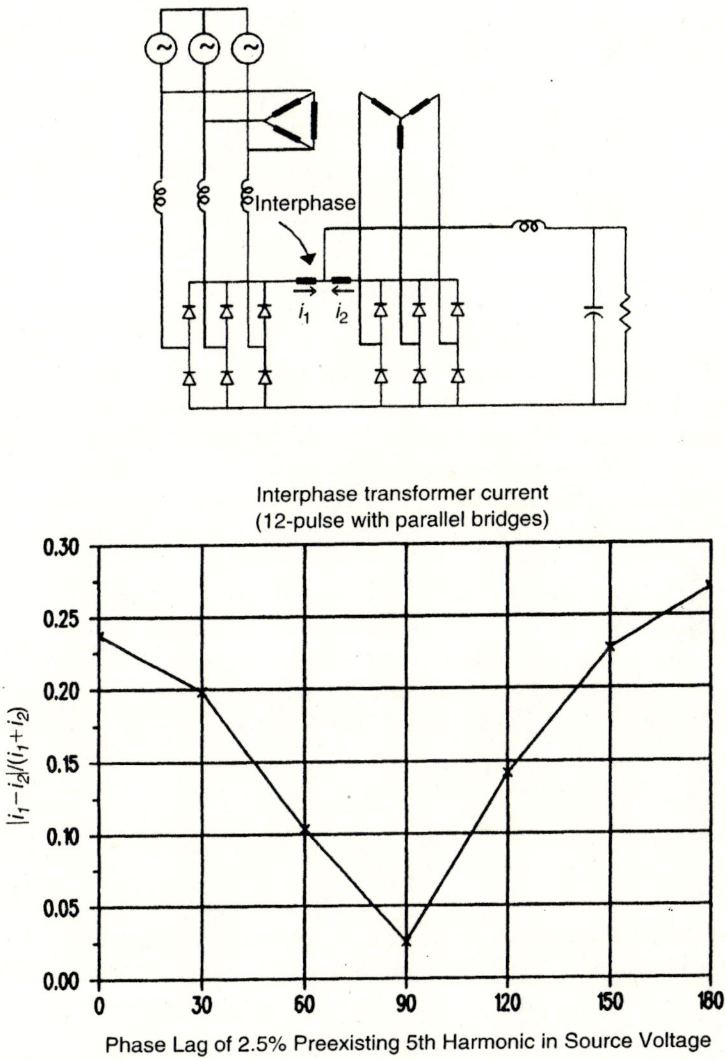

Figure 2-6 Effects of preexisting harmonic voltage on current balance in 12-pulse converter.

power system conditions. The 7th harmonic causes even further unbalance if it has the same phase shift ϕ.

In practice the harmonic voltage phase angle is difficult to define, because it is a function of converter design and system parameters. The precise phase relationships of various harmonics are never known exactly. In one system measurement the 5th harmonic of voltage was found to be nearly in phase with the fundamental—the worst case for multipulse systems using paralleled 6-pulse converters fed from delta/wye connections.

Compensation for dc unbalance can improve performance in some multipulse circuits. This compensation could be by appropriate dc offset injection, using a small high-frequency converter, or by phase control when the converters use thyristors. In the latter case only a fraction of a degree phase difference may be necessary, a design challenge. Another possibility is the use of a harmonic blocking transformer as described in Chapter 6, Section 6.4.

2.6.4 Different Circuit Topologies

The effects of different power sources on different topologies are discussed in Chapter 9. The circuits described may use either diodes or thyristors but in this book the performance characteristics are first highlighted by focussing upon diode operation. For a given connection the performance in the face of different power sources is affected by the amount of ripple current in the dc circuit. In turn, this is affected by the amount of dc circuit inductance. For example, sensitivity to source voltage unbalance is effected by the ratio of $\omega L_{dc}/R_{dc}$, as illustrated in Figure 7-3.

Where very low harmonic currents are desired and the practical source has characteristics of type 2 power, the dc inductor filter can be selected to give good results according to the practical relationship $\omega L_{dc}/R_{dc} = 0.168$ at full load. When the source approaches type 1 power, or when somewhat higher distortion is tolerable, a value of $\omega L_{dc}/R_{dc} = 0.084$ gives good results in most circuit connections.

2.6.5 Summary of Effects of Power Type

It is very desirable to incorporate the practical imperfections of voltage unbalance and preexisting voltage harmonics in analysis of different converter circuits.

One "standard," defined as source power type 2, incorporates 1% negative-sequence voltage and a preexisting 5th harmonic voltage of 2.5%, with variable phase. If left uncorrected, this type of power can cause significant uncharacteristic 5th harmonic currents in conventional (paralleled converter) 12-pulse systems; significant unbalance may be caused in the two converters as well.

Chapter 3

Multipulse Methods and Transformers

3.1 MULTIPULSE METHODS

The term "multipulse" is not defined very precisely. In principle we could imagine it to be simply more than one pulse. However, by popular usage in the power electronics industry, it has come to mean converters, operating in a three-phase system, providing more than six pulses of dc per cycle.

Multipulse methods involve multiple converters connected so that the harmonics generated by one converter are canceled by harmonics produced by other converters. By these means certain harmonics, related to the number of converters, are eliminated from the power source. Multipulse converters give a simple and effective technique for reducing power electronic converter harmonics. They have been widely used in high-power applications in the electrochemical industry. The expanding use of power converters for adjustable-frequency ac controllers has stimulated development of multipulse methods in lower power ratings, down to 100 HP and less.

To explain the basic operation of multipulse converters it is assumed that the dc circuit is filtered such that any ripple caused by the dc load does not significantly affect the dc current. This is true for passive loads and for most converters feeding dc power to voltage-source inverters. It is less likely to be true for inverter loads of the current-source type, where practical filtering and controls may be insufficient to prevent the dc load ripple from affecting the total ripple. In this case the ac line input current will contain a wide range of harmonics, including subharmonics, which are not integer multiples of the supply frequency. These cannot be canceled by simple multipulse methods.

Multipulse systems have two major advantages. They are achieved simultaneously and are

1. Reduction of ac input line current harmonics
2. Reduction of dc output voltage ripple

Reduction of ac input line current harmonics is important as regards the impact the converter has on the power system. Also, it may be essential to meet harmonic current standards.

For converters used in voltage-source variable-frequency drive (VFD) systems, the converter dc output voltage ripple mainly has practical importance for its affect on the design of the dc circuit filter inductance. The eventual dc voltage fed to the inverter is usually smoothed by a large dc electrolytic capacitor.

Multipulse methods are characterized by the use of multiple converters, or multiple semiconductor devices, with a common dc load. Phase-shifting transformers are an essential ingredient and provide the mechanism for cancellation of harmonic current pairs, for example, the 5th and 7th harmonics, or the 11th and 13th, and so on.

By way of introduction to multipulse methods, we can start with multiple converters but with separate dc loads. Consider the circuit shown in Figure 3-1. Here two separate loads are fed from two converters, each having its own supply transformer. One converter bridge is fed through a delta/wye transformer that produces a three-phase set of secondary voltages shifted by 30° with respect to the primary voltage. The other converter bridge is fed by secondary voltages, from the delta/delta transformer, which have no phase shift.

Ideally, the fundamental current of each converter will be in the same phase relationship with respect to the system voltage, that is, zero phase shift. However, some of the harmonic currents are differently phased because of the transformer action as described in detail in Section 3.1.1. In Figure 3-1 these phase relationships are indicated in the expressions for ac currents fed to each inverter. Derivation of these series is given in the appendix.

Due to the phase relationships it is seen that some of the currents in one bridge are in antiphase with those in the other. Thus, we can say that some of the harmonic currents required in one converter are supplied by another. If the converter loads are equal, certain harmonics are eliminated from the supply.

For example, in Figure 3-1 with appropriate amplitudes for i_1 and i_2, the 5th and 7th harmonics can be canceled and the system will "see" the equivalent of a 12-pulse load.

In practical situations the loads will not be precisely balanced. However, this technique provides a means of achieving a limited amount of system harmonic current reduction when multiple converter loads are on a power system. The method is analyzed for 12-, 18-, and 24-pulse arrangements in Chapter 8, Section 8.6, where it is described as multipulse by "phase-staggered" converters.

The dc loads in Figure 3-1 can be directly interconnected to ensure equal loading under all conditions. However, to enable the converters to operate independently of each other and to maintain semiconductor device conduction at 120°, it is necessary to include interphase transformers. These transformers are discussed in Chapter 6. With these in place, the load can always be balanced ex-

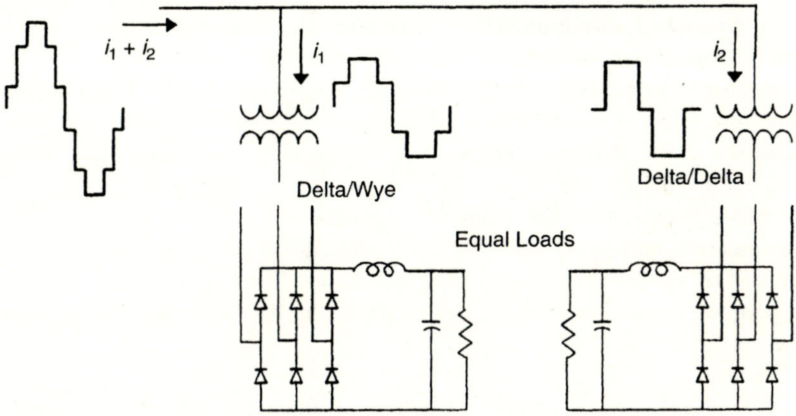

$$i_1(\omega t)\,\alpha\,\frac{2\sqrt{3}}{\pi}\left(\cos\omega t-\frac{\cos 5\omega t}{5}+\frac{\cos 7\omega t}{7}-\frac{\cos 11\omega t}{11}\cdots\right)$$

$$i_2(\omega t)\,\alpha\,\frac{2\sqrt{3}}{\pi}\left(\cos\omega t+\frac{\cos 5\omega t}{5}-\frac{\cos 7\omega t}{7}-\frac{\cos 11\omega t}{11}\cdots\right)$$

Figure 3-1 Two separate 6-pulse converters combine to draw 12-pulse current from the power system.

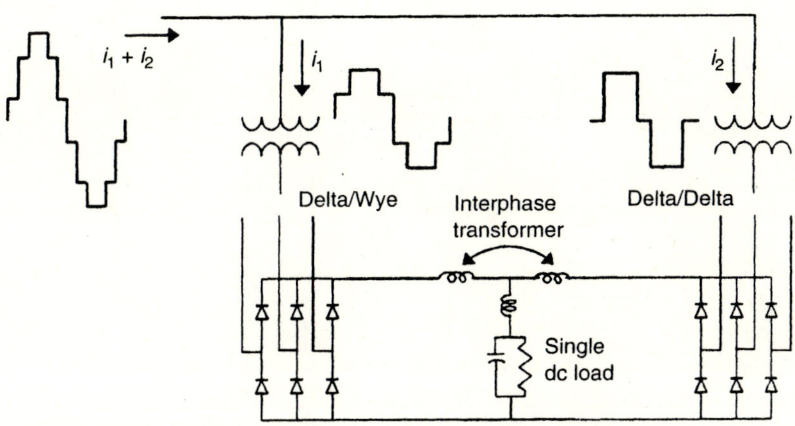

Figure 3-2 Two 6-pulse converters combine within a single equipment and single dc load for continuous 12-pulse operation.

cept for practical limitations. Thus, in Figure 3–2, 12-pulse is obtained with a single ac-to-dc power converter.

An 18-pulse arrangement that parallels three 6-pulse converter bridges, in conjunction with an interphase transformer, is shown in Figure 3-3. Specific transformer connections are not shown on this drawing because the required phase shift, in this case ±20°, can be obtained in many different ways.

Another way to develop multipulse operation is shown in Figure 3–4. This drawing shows paralleling methods deploying multiple, 3–pulse, diode converters. Idealized harmonic orders and amplitudes for line current harmonics and dc voltage ripple are included in the figure. These highlight the advantages of multipulse performance.

The amplitude of each harmonic of input current always varies inversely with its frequency. Thus, with higher pulse numbers that eliminate lower-frequency harmonics, the harmonics present are of smaller amplitude: a very desirable result.

The relationship for individual output ripple voltages is much more complex [2]. However, the peak-to-peak ripple voltage, which gives a "feel" for the quality of the dc output, is readily determined. It is included in Figure 3-4.

If we consider the basic theoretical performance obtained with ideal components and power systems, then the characteristics of multipulse converters fed from phase-shifted power supplies are universal. That is, all ideal multipulse circuits with the same pulse number have the same characteristic harmonic performance.

Specific multipulse connections may include series or parallel arrangements of converters, with 120° conduction, or they may have reduced device conduction angles. In Chapter 9 it is demonstrated that practical variations in compo-

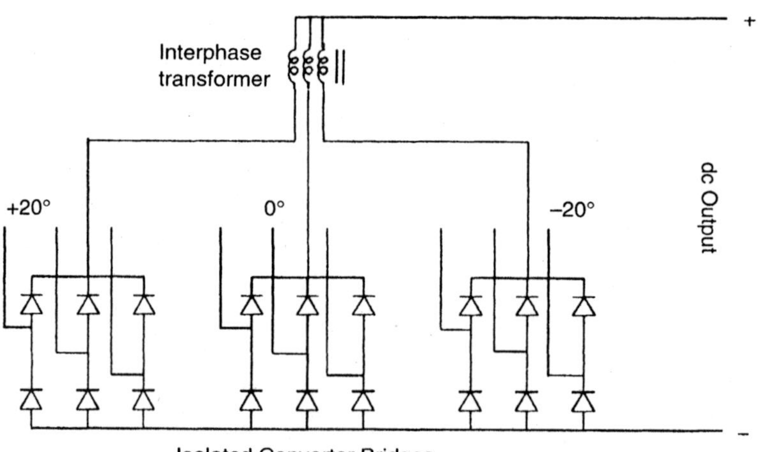

Figure 3-3 Three 6-pulse circuits combined into a single dc load for continuous 18-pulse operation.

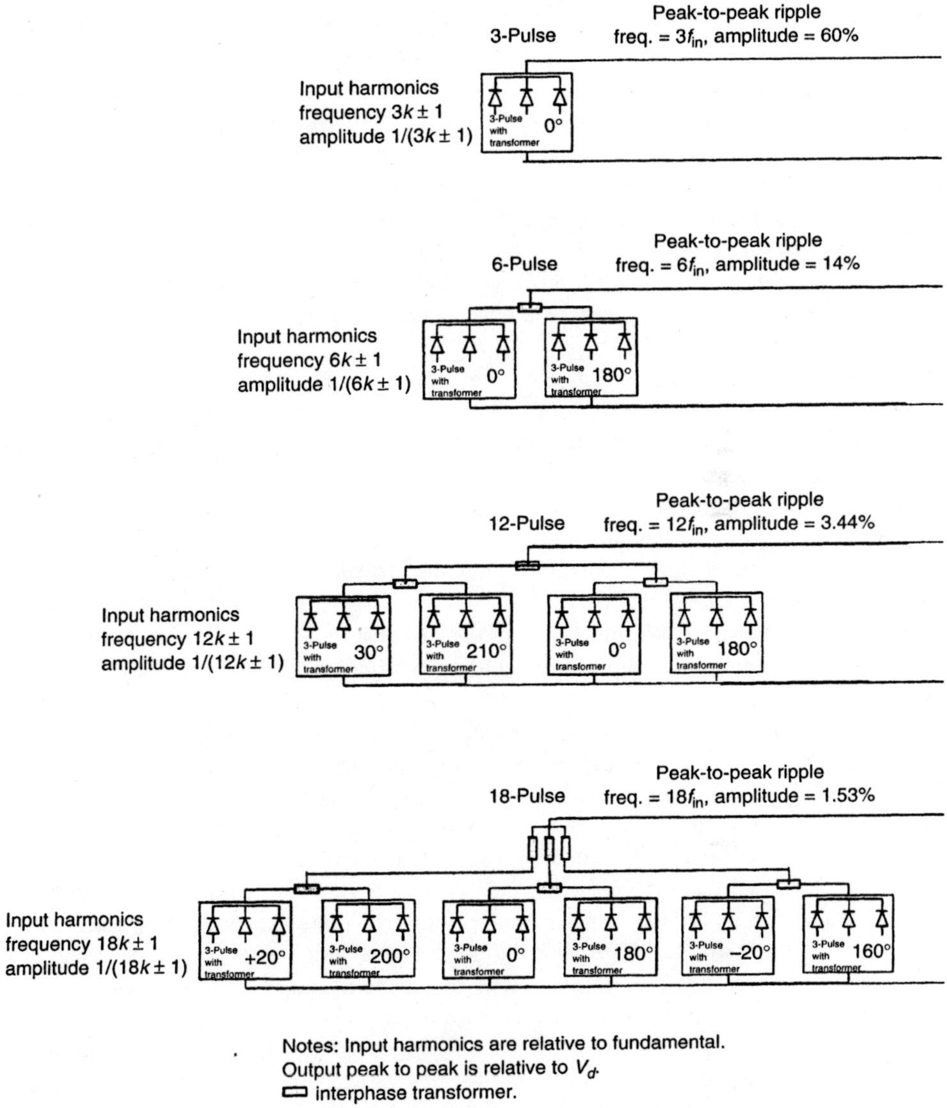

Figure 3-4 Topologies for multipulse connections using 3-pulse diode converters as a "building block."

nent and power source features are more readily tolerated by some multipulse topologies than others.

The three-phase converter bridge topology, deployed in Figures 3-2 and 3-3, is a popular connection. It is an excellent building block for multipulse converters, and produces six pulses of dc output voltage per period of the supply. Characteristic current harmonics have frequencies defined by $(6k \pm 1)$, where k is any positive integer. With a quantity of m 6–pulse converters properly combined, the characteristic current harmonics in the three-phase power source will have frequencies of $(6km \pm 1)$ with amplitudes of $1/(6km \pm 1)$. Thus, three converter bridges appropriately connected, as in Figure 3-3, will cause the supply to have characteristic current harmonics of $(18k \pm 1)$. In this event the predominant current harmonic will be that of the 17th with an amplitude of only 1/17, that is, 5.88%. When the harmonics are reduced to this extent only a small amount of additional filtering is needed for the ac line current to approach a sinusoidal waveform.

The examples given so far have achieved a higher pulse number by direct paralleling of lower-pulse-number converters. This technique is especially popular for very-high-current converters where the current is such that it is necessary to parallel semiconductor devices anyway. As noted previously, an interphase transformer is included when the circuits are paralleled. The interphase transformer absorbs instantaneous voltage differences between the converters and allows individual device current conduction to remain at 120°. This gives good utilization of the device current rating.

Interphase transformers are not easy to design. Schaeffer [1] provides a very helpful discussion of the subject, and in this book the principles are reviewed in Chapter 6.

There are other paralleled converter topologies that provide multipulse performance but do not require interphase transformers. In these, the converters are allowed to influence each other, and the semiconductor device conduction angle is not constrained to be 120°. This increases device and transformer root-mean-square (rms) currents but may be a good trade-off in many cases.

Series converter connections also provide a means for multipulse operation without the use of interphase transformers. In this case the device conduction can be maintained at 120°, but unless the application requires series devices to support the voltage, extra losses will be caused because the total device forward voltage drop is increased.

These circuit variations are discussed in Chapters 4, 5, and 6 in conjunction with the transformers needed to implement them. However, as previously stated, the dc output and ac input harmonic relationships are universal, independent of the particular topology.

Elimination of interphase transformers is particularly desirable when there are preexisting harmonic voltages in the three-phase power source. This is because preexisting source harmonic voltages cause changes in the dc output, which greatly complicates interphase transformer design. This characteristic is analyzed in Chapter 2 and practical examples of the effects are shown in Figure 2-6.

3.1.1 How Harmonics Are Canceled

For harmonic current reduction the multiple converters are fed from phase-shifting transformers. Because a three-phase input is available, we have a simple means of obtaining phase shift by adding appropriate segments of voltage in a transformer. This sort of "cut and paste" of the voltage vectors can be described as phase splitting.

The resulting phase shift obtained must be appropriate for the number of converters. In general, the minimum phase shift required for cancellation of current harmonics in converters with 6-pulse waveforms is

$$\text{minimum phase shift}\,^\circ = \frac{60}{\text{number of converters}} \tag{3.1}$$

In many multipulse circuits the individual harmonic currents of each bridge converter remain the same. The transformer(s) simply allow the harmonic currents required by one bridge to be supplied by another. The fact that negative-sequence voltages and currents are shifted in the opposite sense to positive-sequence values also provides a mechanism to cancel harmonics in pairs.

This is a very fortunate result, and it is instructive to reason why the harmonic currents cancel in the case of two converters fed from supplies with a $+30°$ phase shift.

There is no provision for energy storage (and ideally no power loss) as current is transferred through any transformer. Thus the fundamental component of current at the input must be in the same phase with respect to the input voltage as the output current is with respect to the output voltage. Referred to the phase-shifting transformer input, the fundamental output current has to be moved through the phase shift angle of Φ. Harmonic currents will also be moved through an angle of Φ degrees as they pass through the transformer, either $+\Phi$ or $-\Phi$ depending on sequence. This must be so because the transformer is assumed to operate equally well at any frequency.

The phase angle of the harmonic voltage or current is defined relative to the harmonic frequency at the instant when the fundamental vector is in the reference position. For example, in Figure 2-1, the 5th harmonic is said to be at a phase angle of $-30°$.

Assume the fundamental input vector is moved through an angle of $-\Phi$ in a phase-shifting transformer to produce a lagging output vector. The 5th harmonic in the input will be shifted by an angle of $+\Phi$ in the output. This is because the 5th harmonic is a negative-sequence vector. At a later time when $\omega t = \Phi$, the output fundamental vector will reach the reference position of zero. At this moment in time the 5th harmonic will have moved through a further angle of $+5\Phi$ due to the elapsed time. Thus, the 5th harmonic is at an angle of 6Φ. We can consider the 7th harmonic in a similar manner. Because it is positive sequence it will undergo a phase shift of $-\Phi$ in the transformer. When the output reaches the reference position, the 7th harmonic will have shifted a further $+7\Phi$. Thus, it too will be at an angle of 6Φ with respect to the output.

Both 5th and 7th harmonics can be moved to have an opposite input to output phase position by choosing $\Phi = 30°$. Thus, by feeding two converters with a phase displacement of 30° we can eliminate the 5th and 7th harmonics of current. One way to achieve this is to feed one converter directly from the line and the other from a delta/wye transformer.

For the 5th harmonic vectors the results are illustrated in Figure 3-5. The results apply equally for current or voltage vectors. Similar arguments apply to some of the other harmonics, and it is determined that the 5th, 7th, and in general $6(2k - 1) \pm 1$ harmonics, where k is any positive integer, all cancel when Φ is 30°.

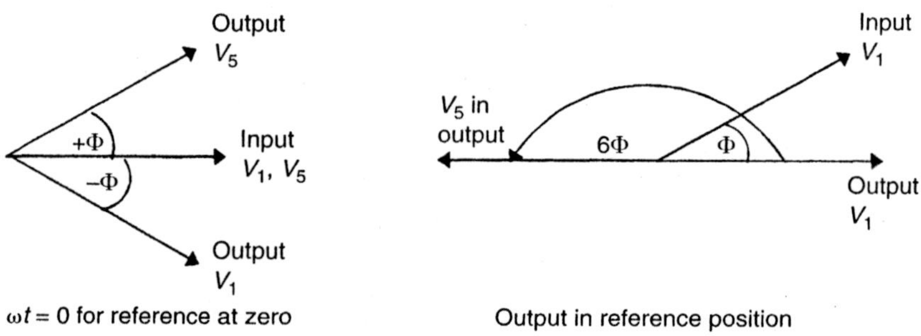

Figure 3-5 Showing that a transformer with fundamental shift of Φ moves 5th harmonic of output by 6Φ.

The discussion so far has centered on paralleled converters, but identical arguments apply to series connection of converters. In another approach for multipulse, the transformers are used to provide a higher number of phases for the converter connection. For example, in a nine-phase rectifier bridge, a nine-phase supply is required. For this the voltage vectors are separated by 40°. This amount of phase shift is twice what is required for three parallel connected 6-pulse converters. However, we shall find that some very excellent results are attained by the nine-phase bridge method.

Whatever technique is used for multipulse converters, an understanding of phase-shifting transformers is required. Before introducing specific transformer connections, a review of some basics should be helpful.

3.2 MULTIPULSE TRANSFORMER BASICS

Transformer practical imperfections such as magnetizing current and copper losses are ignored in the basic design but must be included when system effects are evaluated. This is particularly important when performing energy efficiency calculations.

In practical transformers there is imperfect flux coupling which causes the transformers to exhibit a leakage inductance. Sometimes this may be helpful, as in reducing high-frequency harmonic currents. In other cases it causes difficulties such as asymmetries and excessive voltage drop due to commutation overlap. When reduced leakage inductance is required, bifilar windings can often reduce leakage by a factor of five times or more.

Ignoring leakage inductance when establishing basic phase-shift criteria and winding currents is usually permissible; however, leakage inductance must be considered in the final practical design. This is not an easy calculation. An elegant approach based on stored magnetic energy is described in [1]. Leakage inductance can be a major factor in the performance of transformer designs in excess of 100 kVA.

Transformer windings possess a quality of "starts" and "finishes" which relate to the direction of flux produced for a given polarity of current.

In the development of multiwinding magnetic amplifiers, it was customary to mark the "start" of a winding with a small dot. This technique is also used for transformers. Another method is when windings on the same phase are drawn parallel to each other with the same start or finish polarity being at the end of the winding.

Examples of the two methods of polarity marking are shown in Figure 3-6. In a more complicated transformer arrangement, such as that of Figure 3-7, a combination of methods may be helpful. If the three phases *A, B, C* comprised three single-phase transformers, the "dot" convention would not be necessary.

By using the "dot" convention in Figure 3-7, the windings can be correctly connected for a three-phase, three-limb core. In this case, current in a particular direction produces the same flux direction relative to the core that it surrounds.

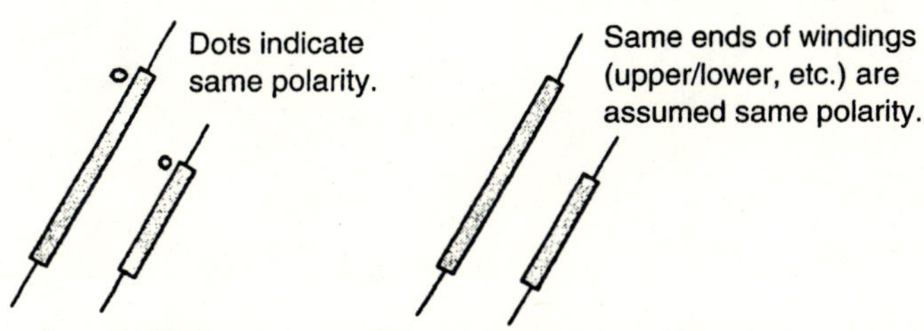

Figure 3-6 Interpretation of transformer polarity.

3.2.1 Determining Phase Shift

Vector diagrams can be very helpful in analyzing various winding configurations. With appropriate nomenclature a physical representation of the windings can conform to the vector diagram. The result is a geometrical figure that, with some simple trigonometry, enables voltages and phase angles to be calculated.

The method requires that all vectors are drawn with the same positive potential definition and that each vector touches another. Figure 3-7 clarifies this technique.

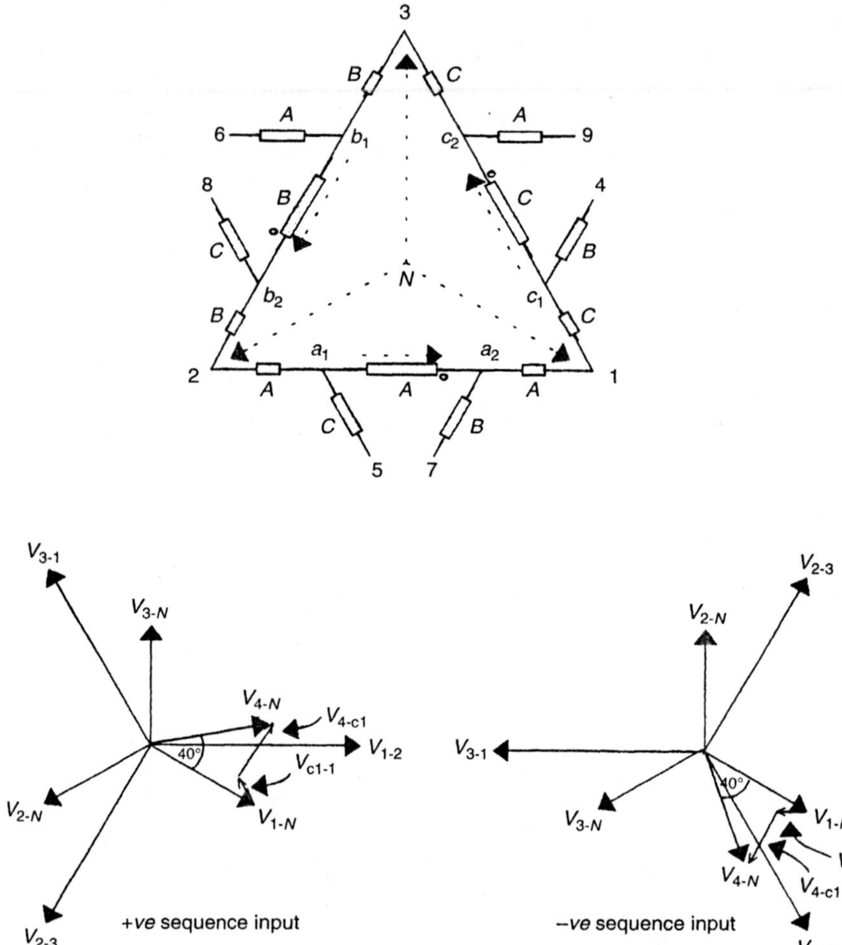

Figure 3-7 Schematic and vectors for a three-phase differential-delta connection with phase-splitting transformer.

3.2.2 Discussion of Vector Representation

The transformer connection shown in Figure 3-7 has been used to generate nine equal amplitude voltage vectors with a 40° displacement. (See Chapter 5, Section 5.4.) The drawing is given here to illustrate the use of geometrical properties to determine required phase shifts. Three phases *A, B, C* are shown, and windings (blocks) on the same phase are marked accordingly. There are five windings per phase. The power connections are made to points 1, 2, and 3. In the left-hand side vector drawing, the sequence of supply voltages is positive, that is, in the order 1, 2, 3. Line-to-neutral (virtual neutral) voltage vectors representing these have been drawn dashed. An arrowhead indicates the positive direction. To draw arrows alongside the windings indicating the positive direction of voltage in those windings, we can start with vector $a_2 - a_1$, which is a fraction of $V_{1\text{-}2}$, and then proceed accordingly with phases *B* and *C*. For example, to determine the voltage $V_{4\text{-}N}$ that is required to be at 40° with respect to $V_{1\text{-}N}$, we can simply add up the appropriate vectors. An example is

$$V_{4-N} = V_{1-N} + V_{c1-1} + V_{4-c1}$$

The vector $V_{4\text{-}c1}$ is worthy of a little extra discussion. It is chosen to ensure that the resultant is a simple vector sum. By this means the geometrical properties are clearer. $V_{4\text{-}c1}$ is developed in the phase *B* winding which is supplied with a vector voltage $V_{2\text{-}3}$. For positive sequence input, this would be shown as a vector 120° lagging on $V_{1\text{-}2}$. Thus the positive direction of the voltage vector on $c_{1\text{-}4}$ would consist of an arrow with the head on c_1 and tail on 4. By defining the vector as $V_{4\text{-}c1}$, we are doing the same thing as adding $-V_{c1\text{-}4}$. This keeps the resultant vector as a summation of positive vectors. In my opinion this facilitates application of trigonometry to determine phase angles. If the vectors are simply defined mathematically, the same result ensues.

If the sequence of the input phases is changed to 1, 3, 2, then the voltage applied at terminal 2 of the transformer leads the line-to-neutral voltage $V_{1\text{-}N}$ by 120°. This is shown in the right-hand vector drawing. The voltage $V_{4\text{-}N}$ is still given by the same algebraic expression, but the vectors are in a different position as shown in the figure. In fact, the phase shift is now –40°. This leads to an important rule for phase-shifting transformers, namely, that *reversing the phase sequence reverses the phase shift.*

3.3 TRANSFORMER KVA RATING

To assist in comparing the cost and losses of different types of phase-shifting transformer, it is helpful to develop a VA rating equivalent to that of a standard double-wound transformer. This equivalent rating is based upon the sum of all products of the sinusoidal equivalent of the mean rectified voltage across the windings (which defines flux) and the relevant rms current. Usually the voltage across the windings will be sinusoidal, or nearly so, but the current will be very dependent upon the converter connection.

Typical double-wound transformers in ratings of 50 to 500 kVA have full-load power losses ranging from 5% to 2% of the kVA rating. By determining the equivalent kVA rating, we can easily determine the range of initial cost and operating power losses. Conventional delta/delta and delta/wye transformers usually have equal primary and secondary kVA ratings, and the effective rating is 0.5 times the combined rating of all windings. Many phase-splitting transformers have windings with unequal ratings; however, an equivalent kVA rating can be calculated in the following manner:

$$\text{equivalent kVA} = 0.5 \times \frac{\sum V_{rms} \times I_{rms}}{1000} \tag{3.2}$$

where V_{rms} is the equivalent sinusoidal rms voltage and I_{rms} is the winding rms current.

All phase-shifting transformer designs must ensure that under fault conditions, such as one bridge being inoperable, the transformer remains protected. This can be achieved by deliberate oversizing of the transformer or by inclusion of thermal trips or both.

3.4 INDUCTOR KVA RATING

In a similar manner it is helpful to make initial sizing and material cost estimates for ac and dc inductors by developing an equivalent kVA transformer rating. Per-unit losses tend to be lower than in transformers, especially for large dc filter inductors. Calculating these losses from the final design is recommended.

The inductor equivalent VA rating will depend upon the product of its equivalent sine wave of voltage V_{eq} and rms current. Thus,

$$\text{inductor equivalent kVA} = (0.5)\, V_{eq} \times I_{rms} \div 1000 \tag{3.3}$$

The rms current can usually be calculated with good accuracy; however, determination of equivalent voltage is more difficult. The equivalent voltage depends upon the available flux capability, which in turn depends upon instantaneous peak current I_{pk} and inductance L. The inductance is easily defined, but determination of I_{pk} must take into consideration any momentary overloads.

Using the centimeter-gram-second (cgs) system of units, we have

$$\text{equivalent voltage} = V_{eq} = 4.44 B_{max} A N f \times 10^{-8} \tag{3.4}$$

where

$B_{\max}$ = instantaneous peak working flux density, in gauss
A = cross sectional area (CSA), in square centimeters
N = number of turns
f = supply frequency (60 Hz)

Using identical units we obtain for the inductance L,

$$L = \text{flux linkages per ampere} = \frac{B_{\max} AN \times 10^{-8}}{I_{pk}} \tag{3.5}$$

Substituting (3.5) and (3.4) into (3.3), we get

$$\text{equivalent inductor kVA} = 0.133\ LI_{rms}I_{pk} \tag{3.6}$$

If the inductors carry sinusoidal currents, then

$$I_{pk} \text{ is equal to } \sqrt{2}\ I_{\max}$$

where $I_{\max}$ is the maximum short-term load.
In this case we get

$$\text{equivalent inductor kVA (sine wave)} = 0.188\ LI_{rms}I_{\max} \tag{3.7}$$

There may be additional features of inductor design, such as dc bias and higher-frequency fluxes, that affect the final result, but (3.6) and (3.7) give useful practical guides.

3.5 SUMMARY OF PHASE-SHIFTING TRANSFORMERS

- If the phase shift is $+\Phi$ for positive-sequence voltage, it will be $-\Phi$ for negative-sequence voltages.
- Cost and losses of phase-splitting transformers can be compared by deriving an equivalent kVA rating.
- Protection must be provided for loss of converter load.
- Winding polarities can be marked with a "dot" system, or by physical positioning of the winding drawings, or both.
- A physical drawing, in which the windings follow all positive (or negative) vectors, facilitates trigonometric calculation of phase shift angles.

Chapter 4

Double-Wound Multipulse Transformers

4.1 GENERAL

A double-wound transformer is one having dc isolation between input and output windings. The kilovoltampere (kVA) rating of the coils and core parts will always be at least twice the load kVA rating.

For example, a 100-kVA transformer has at least 100 kVA of parts in the input and 100 kVA of parts in the output. This is independent of any turns ratio. In contrast, auto connections have dc conducting paths from input to output and a kVA parts rating that may be less than the load, depending upon connection.

The connections presented here have been verified in computer simulations, and most, but not all, have been tested experimentally. Some connections are the subject of patents. Where available, patent numbers have been given. Other connections are newly presented.

Detailed analyses are given for some of the more popular arrangements; in others only a summary is presented. Basic performance and design formulas assume that the dc current is free of ripple. Unless otherwise noted, the supply phase rotation is assumed positive for the direction 1, 2, 3.

Current waveforms are easily sketched in most cases from the 120° conduction flat-topped waves of current that are characteristic of the inputs to a three-phase rectifier bridge. In other cases it may be helpful to eliminate one of the currents using the summation $\Sigma i = 0$.

4.2 DELTA/WYE

Delta/wye is a very widely used arrangement. Because individual winding voltages and currents are not phase shifted, it is highly efficient as regards utilization of the transformer kVA. The 30° phase shift of voltage occurs simply because of relationships between line-to-line and line-to-neutral voltage.

The characteristics in conjunction with a bridge rectifier load are summarized in Figure 4-1. Note that the windings are shown in a physical arrangement that follows the sequence of the system contiguous vectors.

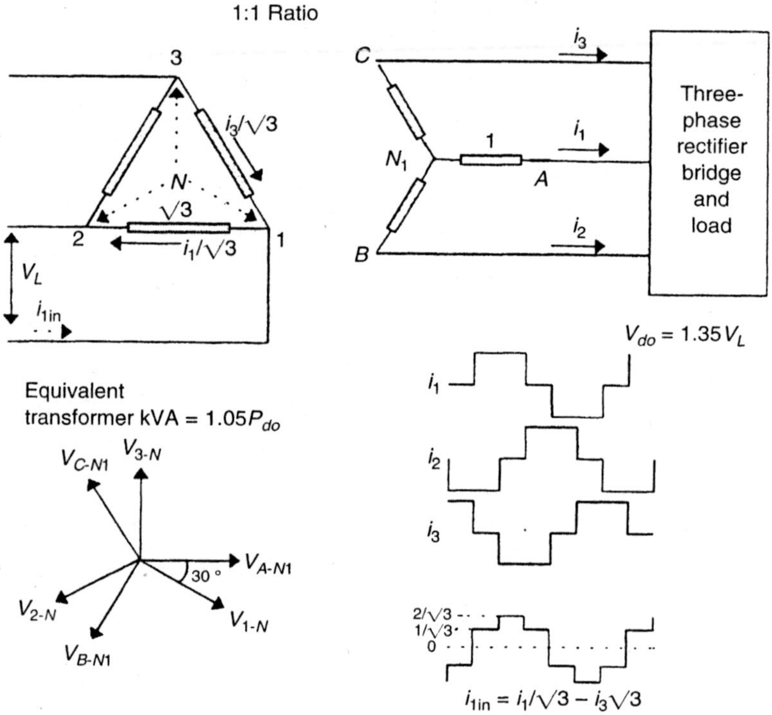

Figure 4-1 The delta/wye connection.

4.3 DELTA ZIGZAG/FORK

The term "zigzag" refers to the manner in which output voltages are formed from segments of different voltage vectors. With a three-phase supply, there are in effect six ways for the voltage vectors to zig and zag. These depend upon selection of one of three angles and one of two polarities (directions). In the fork arrangement the ends of the windings/vectors are left open like a fork.

With two equal amplitude voltage vectors selected, the fork provides a 30° phase shift. This arrangement is sometimes used to supply a 3-diode/SCR, 3-pulse converter connection. For this type of load the fork transformer has the advantage that dc components of current flowing into the 3-pulse connection flow through two equal amplitude but opposite polarity windings. Thus, the net dc ampere-turns, and hence dc flux in the transformer core, sum to zero.

The 3-pulse converter, with three semiconductors, was a popular connection with 3-anode mercury arc rectifiers but no longer finds wide application. This is partly because it represents poor utilization of transformer materials, and it causes significant unwanted line harmonic currents.

Two of the fork secondary winding arrangements can be used with a ±15° phase shift to provide electrically balanced supplies for 12-pulse converters [1]. The two converters could then be connected either in parallel or series.

The basic transformer fork connection is reviewed here in Figure 4-2 with a 1:1 voltage ratio for convenience. The method of connection is clearly satisfactory for any voltage ratio and can be applied to multipulse circuits.

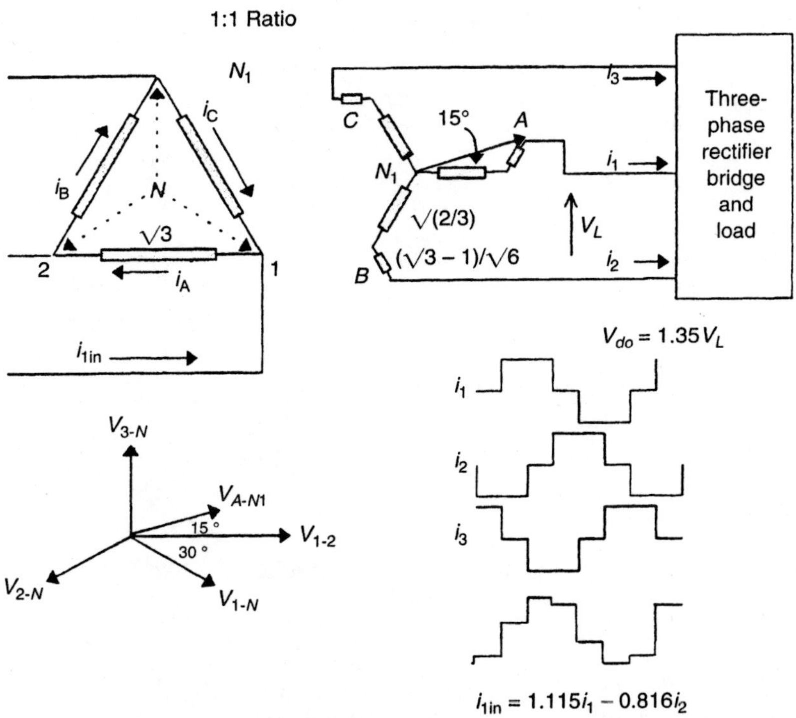

Figure 4-2 Delta/zigzag connection for phase shifting (currents shown for 15° shift).

4.4 DELTA/POLYGON

The secondary windings are connected in a zigzag fashion to form a closed polygon. The root-mean-square (rms) line current is identical to that of the fork connection but the different phase angles change the wave shape. This feature, of identical rms current but different wave shape, can be observed in many of the phase shifting connections. Two electrically identical polygons could be used to provide ±15° phase shift supplies for 12-pulse operation with either series or parallel connection. The polygon topology shown in Figure 4-3 uses a 1:1 ratio for convenience. It is very suitable for other ratios and for multipulse circuits.

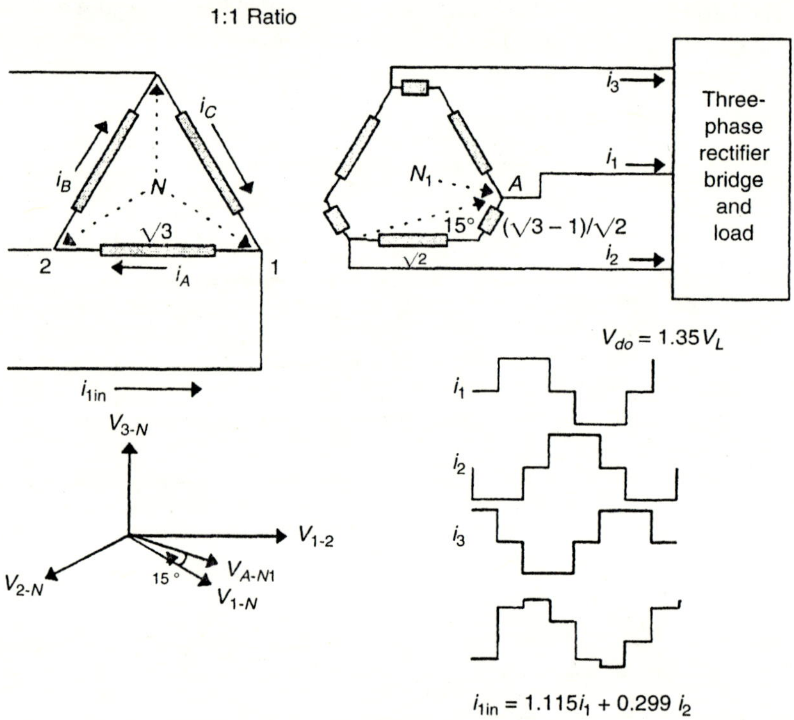

Figure 4-3 The delta/polygon connection for phase shifting (currents shown for 15° shift).

4.4.1 Analysis of Double-Wound Polygon

Analysis of closed polygon connections tends to be more cumbersome than that for open (fork) connections. Therefore, the procedure will be documented for the general case.

Let Φ be the phase shift between polygon output line-to-line voltage V_{Ls} and the delta input line-to-line voltage V_{Lin}. Let the turns/phase on the delta winding be N_p and polygon windings have N_l turns for the long winding and N_s turns for the short winding. Figure 4-4 shows the arrangement for reference.

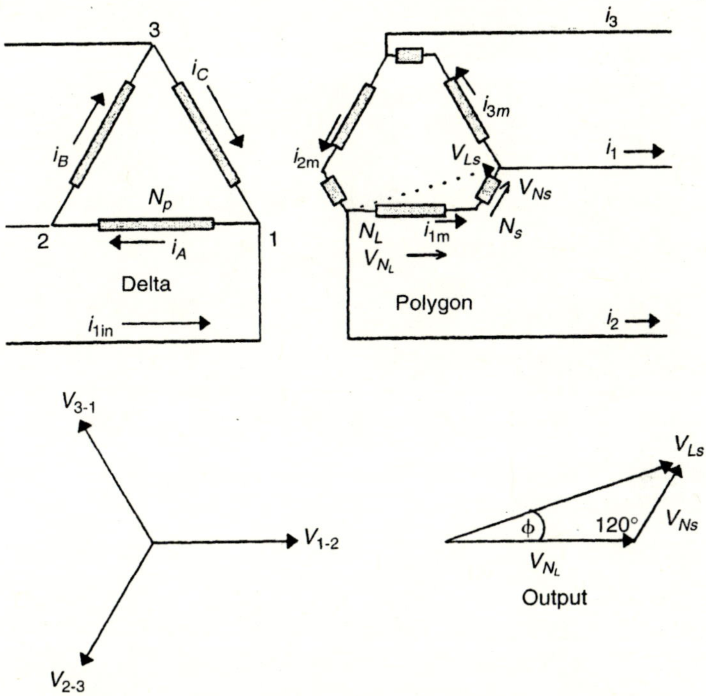

Figure 4-4 Double-wound polygon connection for analysis.

4.4.2 The Polygon Voltages and Phase Shift

From the polygon vector drawing,

$$\frac{V_{Ls}}{\sin(120°)} = \frac{V_{Ns}}{\sin\Phi} = \frac{V_{Nl}}{\sin(60° - \Phi)}$$

Thus,

$$\text{polygon long winding voltage}: V_{Nl} = \left(\frac{2}{\sqrt{3}}\right) V_{Ls} \sin(60° - \Phi) \qquad (4.1)$$

and

$$\text{polygon short winding voltage}: V_{Ns} = \left(\frac{2}{\sqrt{3}}\right) V_{Ls} \sin\Phi \qquad (4.2)$$

From the figure,

$$\tan\Phi = \frac{V_{Ns}\sin(60°)}{V_{Nl+}V_{Ns}\cos(60°)}$$

Thus, for a turns ratio given by $n = V_{Nl}/V_{Ns}$, we obtain

$$\Phi = \tan^{-1}\frac{\sqrt{3}}{(1+2n)} \tag{4.3}$$

4.4.3 The Polygon Winding Currents

$$i_{1m} - i_{3m} = i_1$$

$$i_{2m} - i_{1m} = i_2$$

Subtracting the left sides and right sides of these equations gives

$$i_{1m} - i_{3m} - i_{2m} + i_1 = i_1 - i_2$$

Incorporate $i_{1m} + i_{2m} + i_{3m} = 0$; then,

$$i_{1m} = \frac{i_1 - i_2}{3} \tag{4.4}$$

also

$$i_{2m} = \frac{i_2 - i_3}{3}$$

and

$$i_{3m} = \frac{i_3 - i_1}{3}$$

In general, the converter input currents consist of current pulses with amplitude I_d. They incorporate harmonics of the form ($6k \pm 1$), which include positive- and negative-sequence currents. To calculate the rms current in the winding we can consider these harmonics individually then sum them ($\sqrt{\Sigma i^2}$) to determine the total rms value. The fundamental and rms currents in each converter line are displaced by $\pm 2\pi/3$ depending on their sequence. Thus, each component, and hence the total, is multiplied by

$$\left[\left(1-\cos\pm\frac{2\pi}{3}\right)^2+\left(\sin\pm\frac{2\pi}{3}\right)^2\right]^{1/2}$$

that is, $\sqrt{3}$, as it flows into the secondary windings. The idealized converter line current is a 120° conduction square wave with rms value of $I_d(2/3)^{1/2}$.

Substituting these results into equation (4.4), we obtain

$$\text{secondary winding rms current} = \sqrt{3}\,\frac{1}{3}\sqrt{\frac{2}{3}}\,I_d = 0.471 I_d$$

When load currents i_1, i_2, and i_3, which are characteristic of a three-phase bridge rectifier, are present, we can calculate the line input current to the delta, in the following way. The sum of ampere-turns on the transformer core must be zero.

Thus,

$$i_A N_p - i_{1m} N_l + i_{3m} N_s = 0$$

from which

$$i_A = i_{1m}\left(\frac{N_l}{N_p}\right) - i_{3m}\left(\frac{N_s}{N_p}\right)$$

and, similarly,

$$i_C = i_{3m}\left(\frac{N_l}{N_p}\right) - i_{2m}\left(\frac{N_s}{N_p}\right)$$

The delta winding line input current is given by

$$i_{1in} = i_A - i_C$$

Thus,

$$i_{1in} = \left(\frac{N_l}{N_p}\right)(i_{1m} - i_{3m}) + \left(\frac{N_s}{N_p}\right)(i_{2m} - i_{3m})$$

From Figure 4-4

$$i_{1m} - i_{3m} = i_1 \text{ and } i_{2m} - i_{1m} = i_2$$

thus

$$i_{2m} - i_{3m} = i_1 + i_2$$

Therefore,

$$i_{1in} = i_1\left[\left(\frac{N_l}{N_p}\right) + \left(\frac{N_s}{N_p}\right)\right] + i_2\left(\frac{N_s}{N_p}\right) \tag{4.5}$$

In the example shown in Figure 4-3 where the voltage ratio output/input is 1:1,

$$\left(\frac{N_l}{N_p}\right) = \sqrt{(2/3)} \quad \text{and} \quad \left(\frac{N_s}{N_p}\right) = \left(\frac{\sqrt{3}-1}{\sqrt{6}}\right)$$

The phase shift is governed by the ratio of the secondary turns and is determined to be 15°.

Using these ratios we get, for the line input current,

$$i_{1in} = 1.1153\, i_1 + 0.299\, i_2 \qquad \text{(for 15°)} \tag{4.6}$$

Extending this analysis the polygon winding kVA can be determined in relation to the dc load power. The result is

$$\frac{D.W.\ polygon\ winding\ VA}{V_{do}I_d} = 1.209[\sin(60 - \Phi) + \sin\Phi] \tag{4.7}$$

4.5 DELTA/DELTA/DOUBLE POLYGON

I have not seen this connection described previously, but it represents a straightforward means of providing power for 18-pulse converters. The 6-pulse bridge circuits could be in series or parallel. They are shown here in series to provide tolerance to preexisting power source harmonic voltages. The series connection also eliminates the need for interphase transformers.

The transformer leakage inductances will be an important practical feature in this design. An alternative method using physically separate phase-shifting transformers is also feasible. In principle this connection can be extended to higher pulse numbers by increasing the number of polygon windings and modifying the phase shift accordingly. Practical limitations occur due to reactance effects, efficiency, and the limit on available turns ratios.

Appropriate fork connections could also be employed to provide similar 18-pulse characteristics. However, polygons are easier to use for low-voltage outputs because the higher number of turns facilitates selection of the turns ratios.

Computer-generated current waveforms are shown in Figure 4-5. The input line current is similar to that derived in Figure 8-11 for an 18-pulse phase-staggering method described in Chapter 8.

4.6 DELTA/HEXAGON

This connection is shown in Figure 4-6, page 48, and represents a special case of the polygon. Six output windings per phase are chosen to give a 30° phase shift and to result in a 12-phase output voltage. This output is used to provide two, 30°-shifted, six-phase supplies, each feeding a 6-pulse midpoint converter in which the basic conduction period for the semiconductors is 60°. The term midpoint indicates that the individual output is determined with respect to the neutral of the supply voltages or, as in this case, with respect to a virtual neutral which is at the center of the hexagon.

The two converters are effectively connected in series, and this connection has advantages over many 12-pulse systems that parallel converters.

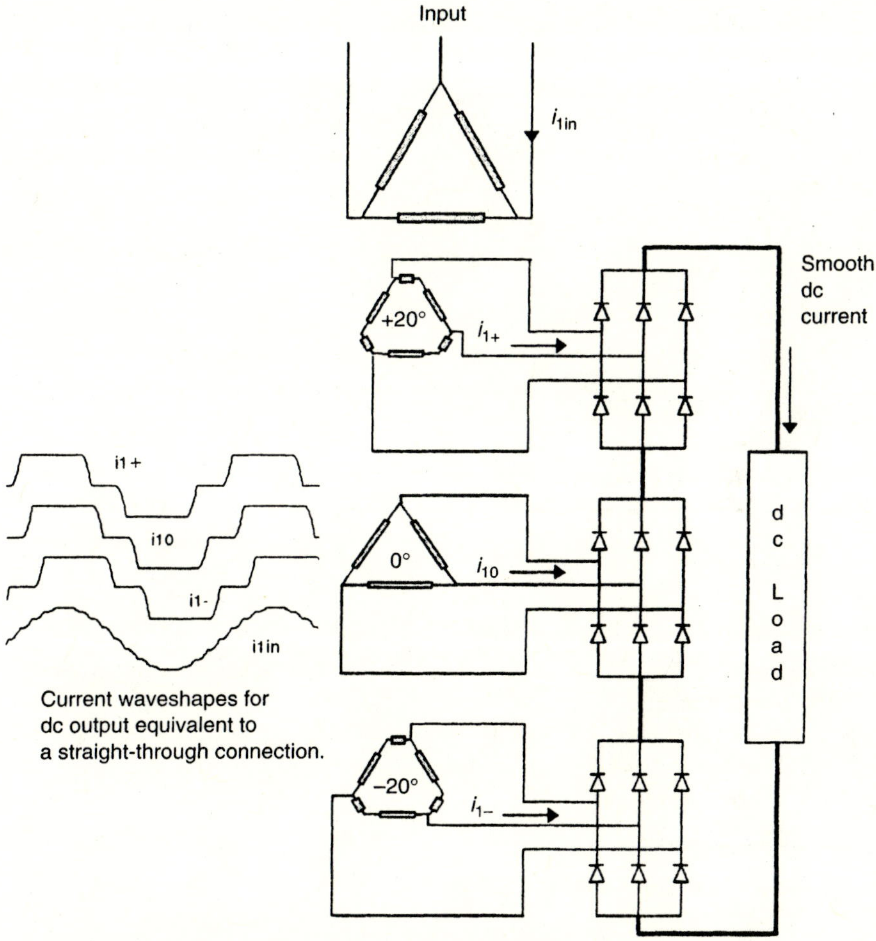

Figure 4-5 Delta/delta/double polygon for 18-pulse.

These advantages include the following:

1. No interphase transformers are needed.
2. It is fairly insensitive to preexisting line voltage harmonics.
3. Additional ac line reactance can be used to very effectively reduce the 11th and higher-order harmonics. For example, with 7% ac line reactance the 11th harmonic of current can be reduced to less than 3% [9].

More information on this connection is given by J. Rosa in U.S. Patent #4,255,784.

An autoconnection adaptation of this connection is given in U.S. Patent #5,148,357. Three-phase power can be fed to terminals such as 1, 9, 5. Alternatively, power can be applied to three apexes of the hexagon to provide a small step down. Another autoconnection that provides voltages for this type of 12-pulse design can be obtained using a unique step-down fork connection. A full description of this method is given in Chapter 9. The circuit has been designated type MC-105.

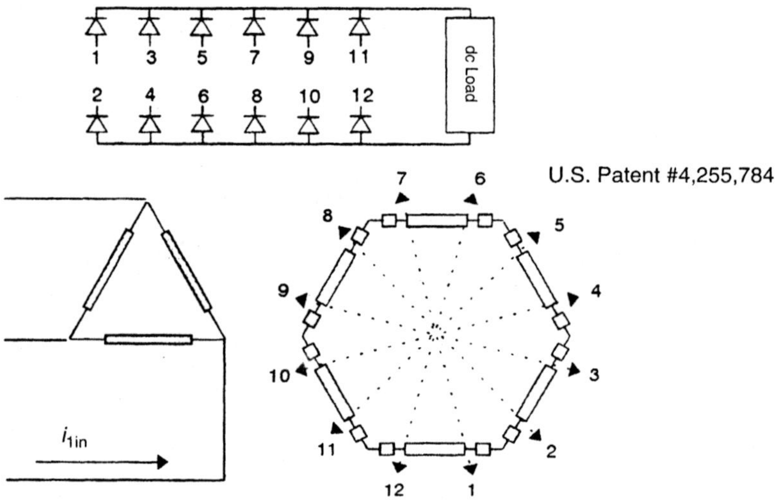

Figure 4-6 The delta hexagon connection for 12-pulse.

Chapter 5

Auto-Wound Transformers

5.1 AUTO-CONNECTED POLYGON

The auto-connected polygon transformer connection, shown in Figure 5-1, is ideally suited to provide phase-shifted power supplies for converters. For a given phase shift, the design is simpler and the parts kVA are lower than the equivalent fork connection.

For example, a 20° phase shift requires only two windings per phase and has a kVA equivalent to a double-wound transformer rated at 31% of the ac load. A fork connection requires three windings per phase and has an equivalent double-wound transformer rating that is about 36% of the ac load.

Because of its high efficiency and fractional kVA rating, the polygon transformer connection is particularly useful as a means to power multiple converter loads and reduce system harmonic currents by means of "phase staggering." This works well in systems with multiple variable frequency drives. Performance of the phase-staggering technique for harmonic mitigation is discussed in Section 8.6.

5.1.1 Design Analysis

Figure 5-1 is used for analysis and derivation of basic design formulas. The per-phase leakage reactance X_L is referred to the long winding. The turns ratio n is defined by the ratio of long winding turns to short winding turns. In deriving basic formulas, practical imperfections such as magnetizing current and leakage inductance are ignored.

5.1.2 Determination of Line Input Current

Referring to Figure 5-1 and noting that the ampere-turns on each limb must sum to zero, we get

$$i_A = \frac{i_3}{n} + \frac{i_B}{n} \tag{5.1}$$

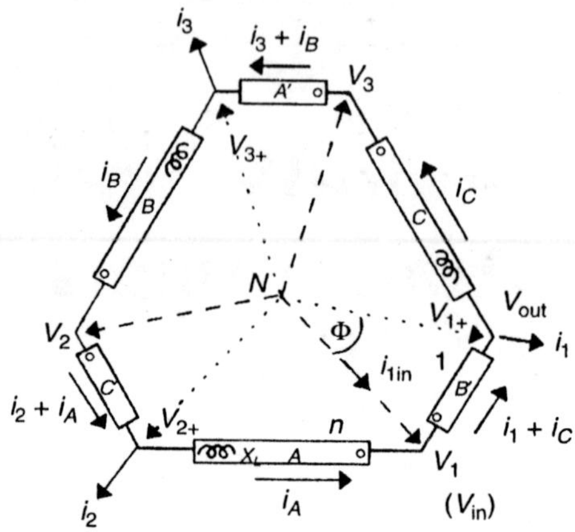

Figure 5-1 Auto-connected polygon transformer.

$$i_B = \frac{i_1}{n} + \frac{i_C}{n} \tag{5.2}$$

$$i_C = \frac{i_2}{n} + \frac{i_A}{n} \tag{5.3}$$

Putting (5.2) in (5.1) gives

$$i_A = \frac{i_3}{n} + \frac{i_1}{n^2} + \frac{i_C}{n^2} \tag{5.4}$$

Putting (5.3) in (5.4) gives

$$i_A = \frac{\left[ni_1 + i_2 + n^2 i_3\right]}{1+n^3} \tag{5.5}$$

Similarly,

$$i_C = \frac{\left[i_1 + n^2 i_2 + ni_3\right]}{1+n^3} \tag{5.6}$$

Now, from Figure 5-1,

$$i_{1in} = i_1 + i_C - i_A \tag{5.7}$$

Putting (5.5) and (5.6) into (5.7) and incorporating ($i_3 = -i_2 - i_1$) yields an equation for input current:

$$i_{1in} = \frac{\left[i_1(n^3 + n^2 - 2n + 2) + i_2(2n^2 - n - 1)\right]}{1 + n^3} \tag{5.8}$$

Voltage and phase angle relationships are readily derived from the figure and design formulas are summarized in the paragraphs that follow.

5.1.3 Summary of Autopolygon Formulas

$V_{\text{out}} = V_{\text{in}}$ at angle Φ, where Φ is positive for positive-sequence and negative for negative-sequence voltages.

$$\text{phase shift}° = \Phi = 2\tan^{-1}\frac{\sqrt{3}}{2n+1} \tag{5.9}$$

$$\text{turns ratio } n = \frac{\sin(60 - \Phi/2)}{\sin(\Phi/2)} \tag{5.10}$$

equivalent double-wound transformer rating

$$= 2.31 \times \text{ac load} \times \sin\left(60 - \frac{\Phi}{2}\right) \times \sin\left(\frac{\Phi}{2}\right)$$

$$= 2.42 \times V_{do} \times I_d \times \sin\left(60 - \frac{\Phi}{2}\right) \times \sin\left(\frac{\Phi}{2}\right) \tag{5.11}$$

If the line voltage is V_L and the line current is I_L, then

$$\text{voltage (long winding)} = \left(\frac{2}{\sqrt{3}}\right) V_L \sin\left(60 - \frac{\Phi}{2}\right) \tag{5.12}$$

$$\text{voltage (short winding)} = \left(\frac{2}{\sqrt{3}}\right) V_L \sin\left(\frac{\Phi}{2}\right) \tag{5.13}$$

$$\text{current (long winding)} = \left(\frac{2}{\sqrt{3}}\right) I_L \sin\left(\frac{\Phi}{2}\right) \tag{5.14}$$

$$\text{current (short winding)} = \left(\frac{2}{\sqrt{3}}\right) I_L \sin\left(60 - \frac{\Phi}{2}\right) \tag{5.15}$$

Equation (5.11) has been plotted in Figure 5-2 to provide a rapid estimate of transformer size.

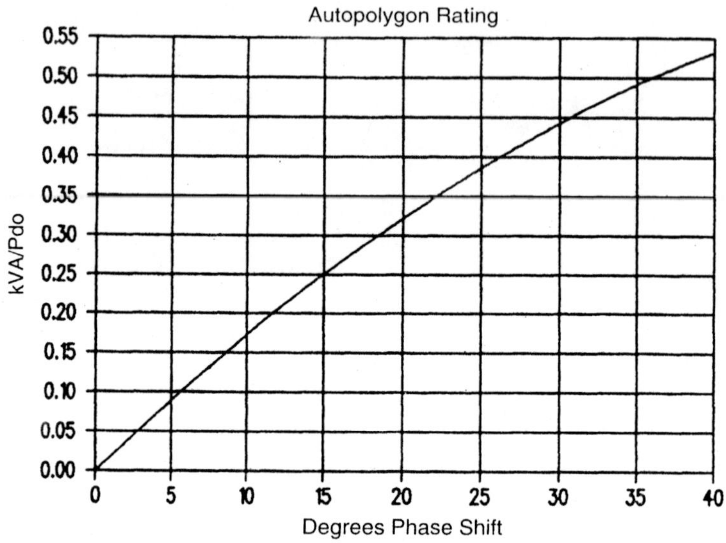

Figure 5-2 Illustrating autopolygon rating.

Equation (5.8) can be used to plot the expected ac line currents when multiple converters are deployed and fed with different phase shifts. The currents i_1, i_2, and i_3 into the converter bridge are typically square wave but with 120° conduction. For positive-sequence voltages applied to the polygon, transformer i_2 will lag 120° on i_1. For negative-sequence voltages, the phase shift will reverse and i_2 will lead i_1. Thus, the shape of the current wave (but not the harmonic content) will depend upon whether the phase shift is positive or negative. An example will make this clear.

Example Assume the auto-connected polygon is required to provide a 20° phase shift. From (5.10) we get $n = 4.411$. Applying this to equation (5.8) results in

$$i_{1in} = 1.134 \times i_1 + 0.386 \times i_2 \tag{5.16}$$

The waveforms defined by this result are plotted in Figure 5-3 on the assumption that i_1 and i_2 are quasi square waves The cases for +20° and −20° are included.

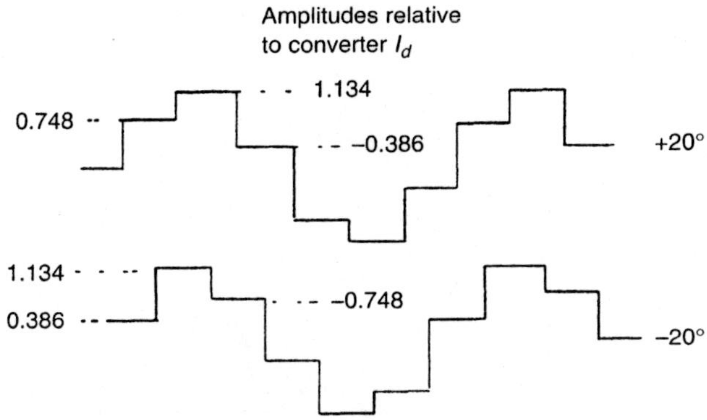

Figure 5-3 Line currents for 20° phase shift polygon.

5.1.4 Commutating Reactance Due to Polygon Transformer Leakage

Like all practical transformers, the polygon transformer exhibits leakage reactance. In classical design calculations, where the dc current is assumed to be free of ripple, the commutating reactance is useful for a number of calculations including line voltage notching. Its value will now be determined.

As a starting point we use the impedance X_L referred to the long winding from a single-phase test. In this test the long and short windings are separated and the short winding connections are bolted together. A low voltage is then applied to the long winding.

During commutation the transformer sees the effect of a short circuit between lines, as shown in Figure 5-4. Applying Thevenin's theorem, the effective impedance "seen" at the output can be obtained by passing a current I_{sc} into the transformer in which the voltage sources are shorted. The effective per-phase impedance is then given by

$$X_{comm} = (1/2)\frac{V_T}{I_{SC}} \tag{5.17}$$

The currents flowing in the mesh connections are controlled by the polygon turns ratio n. For example, a current i_a flowing in the long winding on phase A produces a balancing current, ni_a, flowing in the short winding on phase A. This same current flows in the long winding on phase B and generates a current $n^2 i_a$ in the short winding on phase B. For this current to flow, a voltage drop of $ni_a X_L$ must be developed across the B phase long winding. This causes a voltage drop of $i_a X_L$ across the short winding on phase B.

The analysis continues using Figure 5-4.

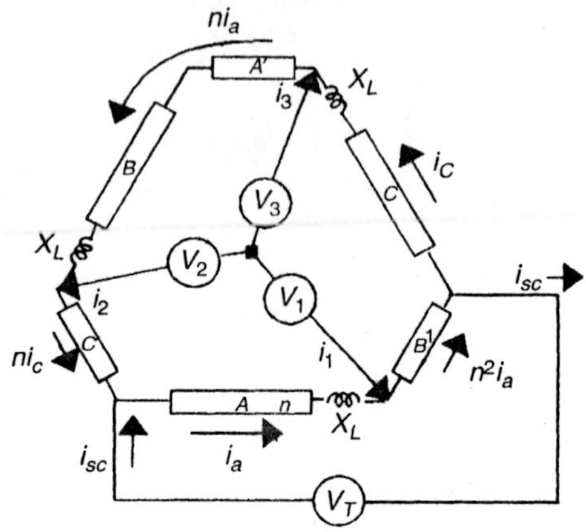

Figure 5-4 Polygon transformer for analysis of X_{comm}. (Note that $V_1 = V_2 = V_3 = 0$.)

From the figure,

$$n^2 i_a - i_{sc} = i_c$$

Also,

$$n i_c + i_{sc} = i_a$$

Combining these, we get

$$i_a = i_{sc} \frac{n-1}{n^3 - 1} \tag{5.18}$$

Also,

$$i_c = i_{sc} \frac{1-n^2}{n^3 - 1} \tag{5.19}$$

Solving for the transformer voltage drops, it is finally shown that the voltage drop across the phase A windings is zero. Thus the voltage V_T developed across the transformer mesh by current i_{sc} is given by

$$V_T = i_a X_L + \frac{n i_a X_L}{n} = 2 i_a X_L \tag{5.20}$$

Combining (5.20) and (5.18) we obtain the result

$$X_{comm} \text{ per phase} = \omega L_{comm} = \frac{X_L}{1+n+n^2} \tag{5.21}$$

To calculate the *idealized* notching volt-seconds, the inductance L_{comm} can be summed with other power source series inductance to get the total inductance L_{Tcomm}.

For a 6-pulse bridge converter fed with a 30° phase shift, we then obtain

$$\text{notch volt-seconds} = \sqrt{3}\, L_{Tcomm}\, I_d \tag{5.22}$$

Further discussion of notching effects is given in Chapter 1.

5.2 AUTO-CONNECTED FORK

As an autoconnection the fork is less effective than the polygon connection when a phase shift is required. Specifically, it requires a larger parts kVA by about 21% for a 30° phase shift. The schematic is shown in Figure 5-5. This connection can be solved in the usual manner by applying the ampere-turns law and Kirchoff's law of current. The derivation is quite lengthy, so only the result is included here.

The expression for the long winding current is shown within the figure. It assumes a turns ratio of n between the long and short windings. The line input current is simply the sum of this current and the known output current into the rectifier bridge.

Although the fork autoconnection is not widely used, it is included here for the sake of completeness and because it is advantageous to be aware of the many connections that can be deployed to produce phase shifts. This knowledge is of practical value when a field harmonics problem is being resolved using the phase-staggering method, for example. Available materials may require that no connection is rejected without evaluation.

$$n = \frac{\sin(60 - \Phi/2)}{\sin(\Phi/2)} \tag{5.23}$$

$$i_A = \frac{i_{L2}(1-n) + i_{L3}(2+n)}{1+n+n^2} \tag{5.24}$$

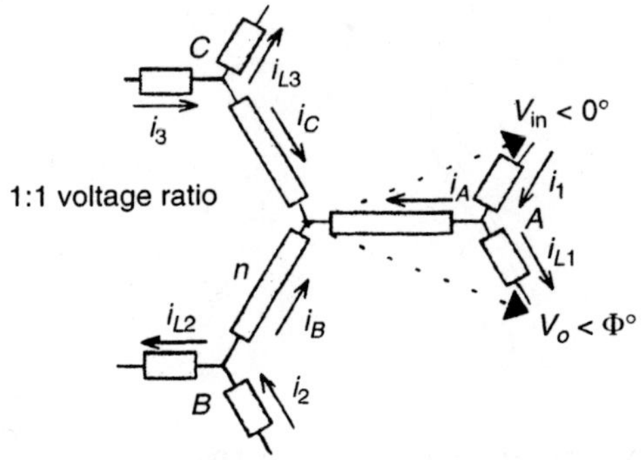

Figure 5-5 Fork connection for a single phase-shift.

5.3 DIFFERENTIAL DELTA CONNECTION

The term "differential" is used here to indicate that some of the transformer windings are only required to carry ampere-turns that are the difference of load current ampere-turns. By these means the transformer kVA rating can be reduced. The practical design must anticipate performance under fault conditions, and this may slightly affect the final rating.

One example of the differential technique was cited as prior art in the 1981 U.S. Patent #4,255,784. The method can be applied to other topologies and is reviewed here in an autoconnection. It often allows the use of a single transformer structure, which helps reduce costs.

Figure 5-6 shows a differential-delta connection in which two sets of output voltage are developed from the three-phase input. The output voltages are higher than the input voltage by a factor of (1/cos Φ/2), where Φ/2 is half the angle between the two output vectors. For analysis the voltage vector diagram is first solved to determine the ratio n of the long winding turns to a short winding. The currents can then be calculated from the six ac currents flowing into the converters.

For a required phase shift of Φ/2 the turns ratio n is given by

$$n = \frac{\sqrt{3}}{\tan(\Phi/2)}$$

In analyzing winding currents it is instructive to observe how readily the rms value of the long winding current is calculated using the current versus time graphs. The elegance of this approach is reinforced if one attempts a complete mathematical solution using, for example, Fourier series expansions.

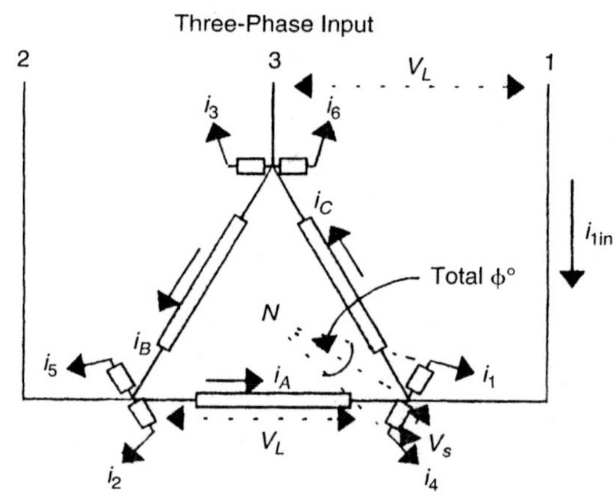

Figure 5-6 Differential-delta phase-splitting transformer, giving a Φ/2 phase shift.

For a 12-pulse converter the total required phase shift is 30°. Thus one converter bridge will be fed with transformer voltages at +15° and the other at −15°. If the total dc load current is I_d, then the currents i_1 and i_4 which feed the two converter bridges will ideally have an amplitude of $I_d/2$. Each converter line current will have a conduction period of 120°, but i_4 will lag i_1 by 30°. Consideration of ampere turn balance on phase B shows that the long winding will carry the difference of ampere turns such that

$$i_B = \frac{i_1 - i_4}{n}$$

This current will have an amplitude of $I_d/2n$. The temporal performance for phase A is shown in Figure 5-7. If the phase displacement between i_1 and i_4 is Φ (degrees), then the rms value of the long winding current is given by

$$i_B = \left(\frac{I_d}{2n}\right)\sqrt{\Phi/90}$$

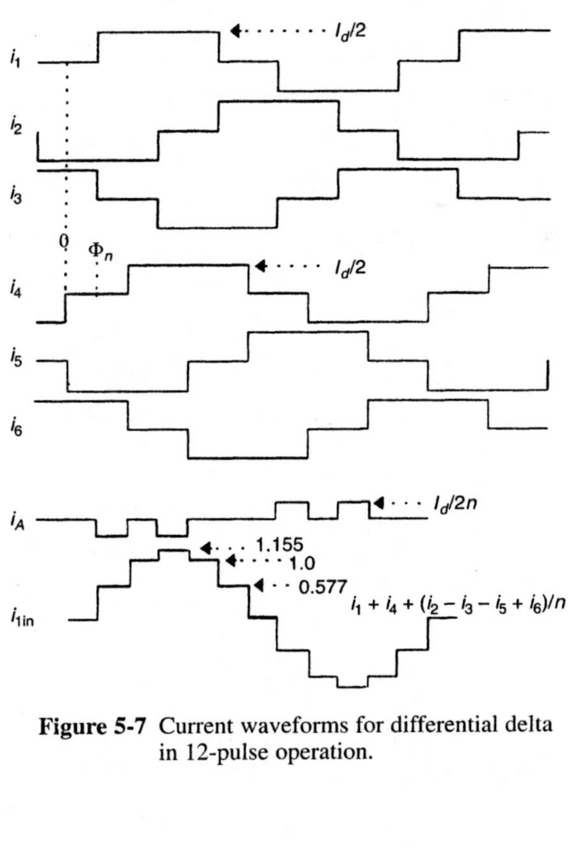

Figure 5-7 Current waveforms for differential delta in 12-pulse operation.

For a 12-pulse arrangement, $\Phi = 30°$ and $n = 6.46$; thus, from this straightforward analysis, we get

$$i_B = 0.045 I_d$$

An analytical result using the spectrum of currents defined by Fourier expansion is useful when knowledge of individual harmonics is required. This will now be derived; however, calculation of rms currents by these means is much more cumbersome. The accuracy is governed by the number of terms considered.

As shown in the appendix, 6-pulse converter ac line currents can be represented by

$$i_n = \frac{2\sqrt{3}}{\pi}\frac{I_d}{2}\left[\sin(\omega t - \Phi_n) - \left(\frac{1}{5}\right)\sin 5(\omega t - \Phi n) - \left(\frac{1}{7}\right)\sin 7(\omega t - \Phi n) + \cdots\right] \tag{5.25}$$

where Φ_n is the angular difference between i_1 and i_n.

For example, the expansions for i_1 and i_4 are obtained by putting Φ_n = zero and $\Phi_n = \pi/6$, respectively.

The difference between any two currents, such as i_1 and i_n, is found using the previous expansion as

$$(i_1 - i_n)\text{rms} = \frac{2\sqrt{6}}{\pi}\frac{I_d}{2}\sqrt{(\sin\Phi_n/2)^2 + (1/5\sin 5\Phi_n/2)^2 + \cdots} \tag{5.26}$$

This series converges quite rapidly. In a 12-pulse arrangement where Φ_n is 30°, considering just the first three terms gives a result within 6% of the actual. The final long winding current in the differential delta transformer is, as shown earlier, equal to $(i_1 - i_4)$/(turns ratio). These somewhat elaborate calculations, using Fourier transforms, highlight the effectiveness of a graphical approach in some cases.

The sum of two bridge converter currents may be useful later in considering multipulse performance. Continuing the development here it is found that

$$i + i_n = \frac{4\sqrt{3}}{\pi}\frac{I_d}{2}\left[\sin\left(\omega t - \frac{\Phi_n}{2}\right)\cos\frac{\Phi_n}{2} - \left(\frac{1}{5}\right)\sin 5\left(\omega t - \frac{\Phi_n}{2}\right)\cos\frac{5\Phi_n}{2} - \left(\frac{1}{7}\right)\sin 7\left(\omega t - \frac{\Phi_n}{2}\right)\cos\frac{7\Phi_n}{2} + \cdots\right] \tag{5.27}$$

where $I_d/2$ is the amplitude of the 120° conduction square wave of current in each converter bridge. From this we get

$$(i_1 + i_n)\text{rms} = \frac{2\sqrt{6}}{\pi}\frac{I_d}{2}\sqrt{(\cos\Phi_n/2)^2 + (1/5\cos 5\Phi_n/2)^2 + \cdots} \tag{5.28}$$

5.4 DIFFERENTIAL DELTA WITH THREE OUTPUTS

In this connection (U.S. Patent #5,124,904), the concept of the differential-delta connection has been extended to provide three equal sets of ac output voltage. It is believed to be optimum in the sense that it has a minimum of kVA parts. The outputs are used to feed two 9-pulse midpoint systems that combine to make an 18-pulse converter.

This connection is amenable to very simple filtering by means of ac line inductance, which includes the transformer leakage inductance. This is because the voltage effecting commutation is much lower than in a 3-pulse system. Therefore, the rate of rise of current at commutation is much reduced, and higher frequencies are readily attenuated. This characteristic is further discussed in Chapter 9, Section 9.2.

Most of the high-frequency filtering occurs because of the transformer leakage inductance. In one practical example the resulting line input currents meet the IEEE Std 519-1992 specification with the largest harmonic of current being less than 1.5%.

A basic schematic is shown in Figure 5-8. In practice an extender winding may be added to the delta to step down the three output voltages. This is because the inherent dc output of an 18-pulse converter, using 9-pulse midpoint groups, is about 18% more than that of a 6-pulse converter.

Due to the proprietary nature of this connection, a full range of formulas will not be given here. Because of the delta characteristics, which require more

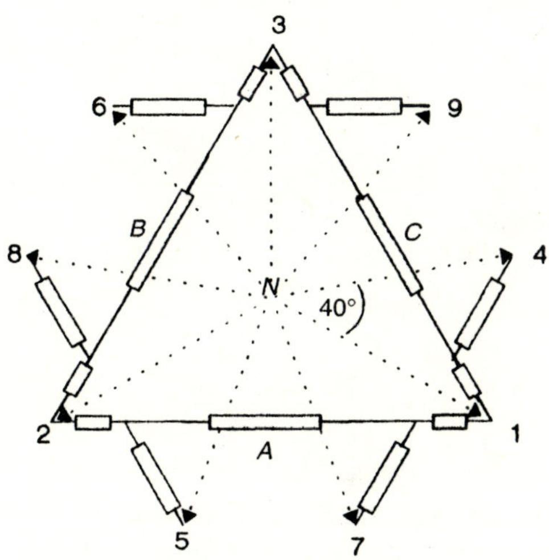

Figure 5-8 Differential delta to give three equal ac outputs for 18-pulse.

turns than a wye connection, there are several turns combinations that allow the phase shift angle to be 40° ± 0.5°. The amplitudes are 1.0 ± 0.5%.

Algebraic solutions for this connection are potentially more cumbersome than those for transformers with two outputs. Digital simulations have proved to be invaluable.

5.5 DIFFERENTIAL FORK WITH THREE OUTPUTS

A general approach for a multiple output fork connection is shown in Figure 5-9. This can provide an alternative means for obtaining three equal sets of three-phase output suitable for feeding a nine-phase bridge converter with 18-pulse characteristics.

Symmetry of the fork connections when a step-down or step-up in voltage is required helps to balance leakage inductance effects. This, in turn, helps balance semiconductor device currents. Also, this symmetry should make it possible to incorporate appropriate leakage inductance to help filter high-frequency harmonic currents. Because of the triplen harmonic currents required in the individual converter currents, a closed delta winding is provided in which the 3rd harmonics of current can circulate.

In computer simulations this connection performs just as satisfactorily as the field-proven differential-delta connection in Figure 5-8. However, Figure 5-10 shows a unique simplification that can be applied in many cases.

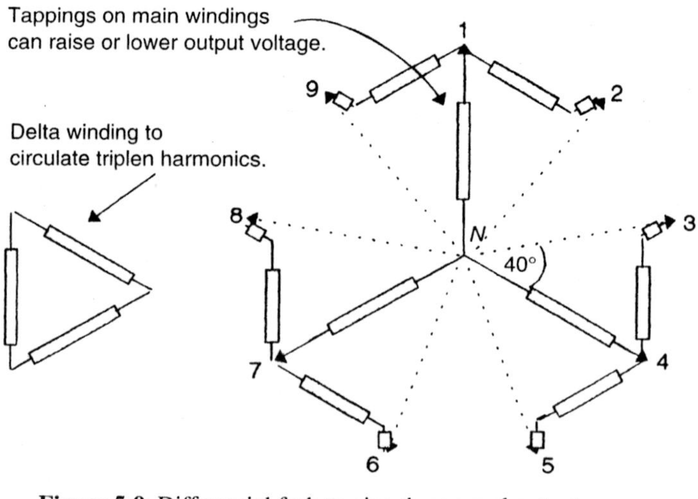

Figure 5-9 Differential fork to give three equal outputs.

Under certain step-down conditions when the ac output voltage is only 87.9% of the supply, the small windings at the ends of the "arms" can be omitted. This is shown in Figure 5-10. This is an important case because now the fork only requires five windings per phase.

The connection in Figure 5-10 has been selected for a detailed analysis in this chapter. The circuit is highlighted in Chapter 9 for computer analysis in the face of practical variations. Specifically, the impact of type 2 power with its unbalance and preexisting harmonic voltages is addressed.

I am not aware of this connection being exposed to the full development process. It is given here as an example of one different way of obtaining a power source for 18-pulse converters.

The results of a preliminary computer simulation are very encouraging. The transformer kVA rating is less than 70% of the dc load power. Viewed from the input the results are excellent, the same as those obtained for the proven differential delta connection.

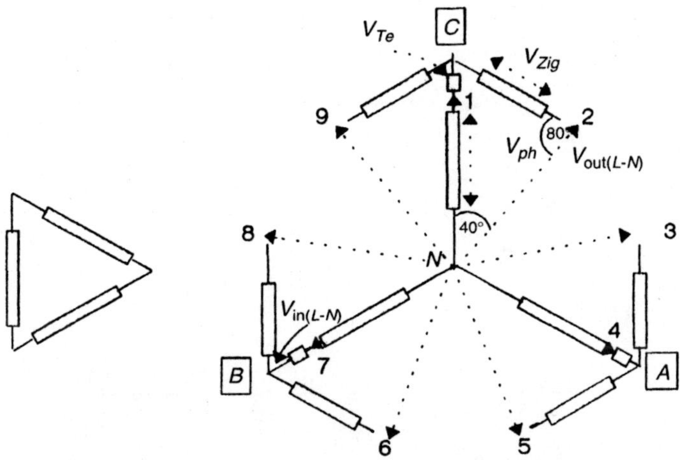

Figure 5-10 Step-down differential fork for 18-pulse converters.

5.5.1 Analysis of Step-Down Fork for 18-Pulse Converters

The schematic used for calculation of voltages is given in Figure 5-10. A three-phase power source of line-to-neutral value $V_{in(L\text{-}N)}$ is applied to the terminals A, B, and C. From the figure, the voltage on the "zig" winding is obtained from

$$\frac{V_{zig}}{\sin 40^\circ} = \frac{V_{in(L-N)}}{\sin 80^\circ}$$

Thus,

$$V_{zig} = 0.653\ V_{in(L\text{-}N)}$$

The voltages on the "teaser" winding V_{Te} and on the phase winding V_{ph} and the output line-to-line voltage $V_{out(L\text{-}N)}$ are determined in a similar manner using the geometrical properties of the figure. Results are summarized as follows.

5.5.2 Voltage Relationships

$$V_{zig} = 0.6527 V_{in(L\text{-}N)} \tag{5.29}$$

$$V_{Te} = 0.1206 V_{in(L\text{-}N)} \tag{5.30}$$

$$V_{out(L\text{-}N)} = V_{ph} = 0.8794 V_{in(L\text{-}N)} \tag{5.31}$$

$$\text{delta winding voltage} = V_{zig}\ \text{(for convenience)}$$

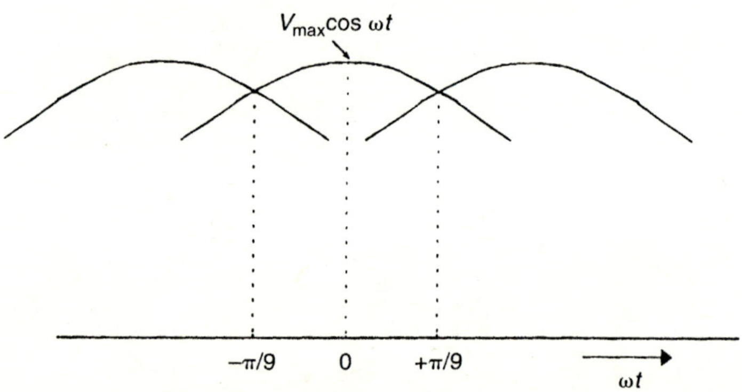

Figure 5-11 DC output voltage of a 9-pulse midpoint group.

From Figure 5-11, we obtain

$$V_{dc} = \frac{9}{2\pi} V_{max} \int_{\omega t=-\pi/9}^{\omega t=\pi/9} \cos\omega t\, d\omega t$$

$$= \frac{9}{\pi} V_{max} \sin\frac{\pi}{9}$$

$$= 0.9798 V_{max} \tag{5.32}$$

Now the total dc output voltage comprises, in effect, two 9-pulse midpoint converters in series; thus,

$$V_{do} = 2 \times 0.9798 \times \sqrt{2} \times V_{out(L\text{-}N)}$$

$$V_{do} = 2.77\ V_{out(L\text{-}N)}$$

$$= 1.6\ V_{out(L\text{-}L)\,(120°)} \tag{5.33}$$

For the special turns ratio in Figure 5-10, we get

$$V_{do} = 1.407\ V_{in(L\text{-}L)} \tag{5.34}$$

The dc output from a straight-through, three-phase converter bridge is 1.35 $V_{in(L\text{-}L)}$. Thus, the special fork connection provides 4.2% more dc output. This extra voltage is tolerable for most converters and helps compensate for voltage drops due to transformer leakage reactance.

5.5.3 Current Relationships

A separate drawing (Figure 5-12 on page 65) is used to solve for the individual winding currents. This schematic defines the currents in the various windings.

The six auxiliary (zig) windings are each assumed to have unity turns. This makes it easier to keep track of the algebra. The long (phase) winding has n_1 turns and the short (teaser) winding has n turns. For convenience in analysis, the delta winding has a single turn, the same as the zig winding. In practice the delta winding can have any convenient number of turns.

In a practical design the individual phase current pulses are heavily dependent upon the per-phase reactance. For example, relatively small reactance values can extend the device conduction period from the idealized value of 40° to values of 60° or more. Because of this characteristic, which results from a small commutating voltage, the currents have a better form factor than those obtained in an idealized analysis. The high-frequency ($18k \pm 1$) harmonic currents are easily filtered by the transformer leakage reactances.

5.5.4 Analysis of Currents

Referring to Figure 5-12,

$$i_a = i_1 - i_A \text{ (Kirchoff's law of current, KLC)}$$

Also,

$$-n_1 i_A + n i_a - i_3 - i_8 - i_D = 0 \text{ (from } \Sigma\, AT = 0)$$

Thus,

$$-(n_1 + n) i_A + n i_1 - i_3 - i_8 - i_D = 0 \tag{5.35}$$

In a similar manner it is determined that

$$-(n_1 + n)i_B + ni_4 - i_2 - i_6 - i_D = 0 \tag{5.36}$$

$$-(n_1 + n)i_C + ni_7 - i_5 - i_9 - i_D = 0 \tag{5.37}$$

From KLC, $-i_C = i_A + i_B$ and substituting this in (5.37), we get

$$(n_1 + n)i_A + (n_1 + n)i_B + ni_7 - i_5 - i_9 - i_D = 0 \tag{5.38}$$

Putting (5.36) in (5.38) gives

$$(n_1 + n)i_A + ni_4 - i_2 - i_6 - i_D - ni_7 - i_5 - i_9 - i_D = 0$$

Thus,

$$(n_1 + n)i_A - i_2 + ni_4 - i_5 - i_6 - ni_7 - i_9 - 2i_D = 0 \tag{5.39}$$

From (5.35)

$$-2i_D = 2(n_1 + n)i_A - 2ni_1 + 2i_3 + 2i_8 \tag{5.40}$$

Put (5.40) in (5.39); then,

$$(n_1 + n)i_A - i_2 + ni_4 - i_5 - i_6 + ni_7 - i_9 + 2(n_1 + n)i_A - 2ni_1 + 2i_3 + 2i_8 = 0$$

from which

$$3(n_1 + n)i_A - 2ni_1 - i_2 + 2i_3 + ni_4 - i_5 - i_6 + ni_7 + 2i_8 - i_9 = 0 \tag{5.41}$$

$$i_A = \frac{2ni_1 + i_2 - 2i_3 - ni_4 + i_5 + i_6 - ni_7 - 2i_8 + i_9}{3(n_1 + n)}$$

In the selected example, to give $\pm 40°$, if the zig winding $= 1.0$, $n_1 = 1.347$, $n = 0.185$, then

$$i_A = \frac{0.37i_1 + i_2 - 2i_3 - 0.185i_4 + i_5 + i_6 - 0.185i_7 - 2i_8 - i_9}{4.596} \tag{5.42}$$

The current in the one-turn delta winding is found by substituting equation (5.41) into (5.35). This yields

$$i_D = \frac{ni_1 - i_2 - i_3 + ni_4 - i_5 - i_6 + ni_7 - i_8 - i_9}{3} \tag{5.43}$$

From Figure 5-12 the input current is given by

$$I_{in} = i_a + i_2 + i_9$$

From Kirchoff's law of current, the teaser winding current i_a is given by

$$i_a = i_1 - i_A$$

Thus,

$$I_{in} = i_1 - i_A + i_2 + i_9 \tag{5.44}$$

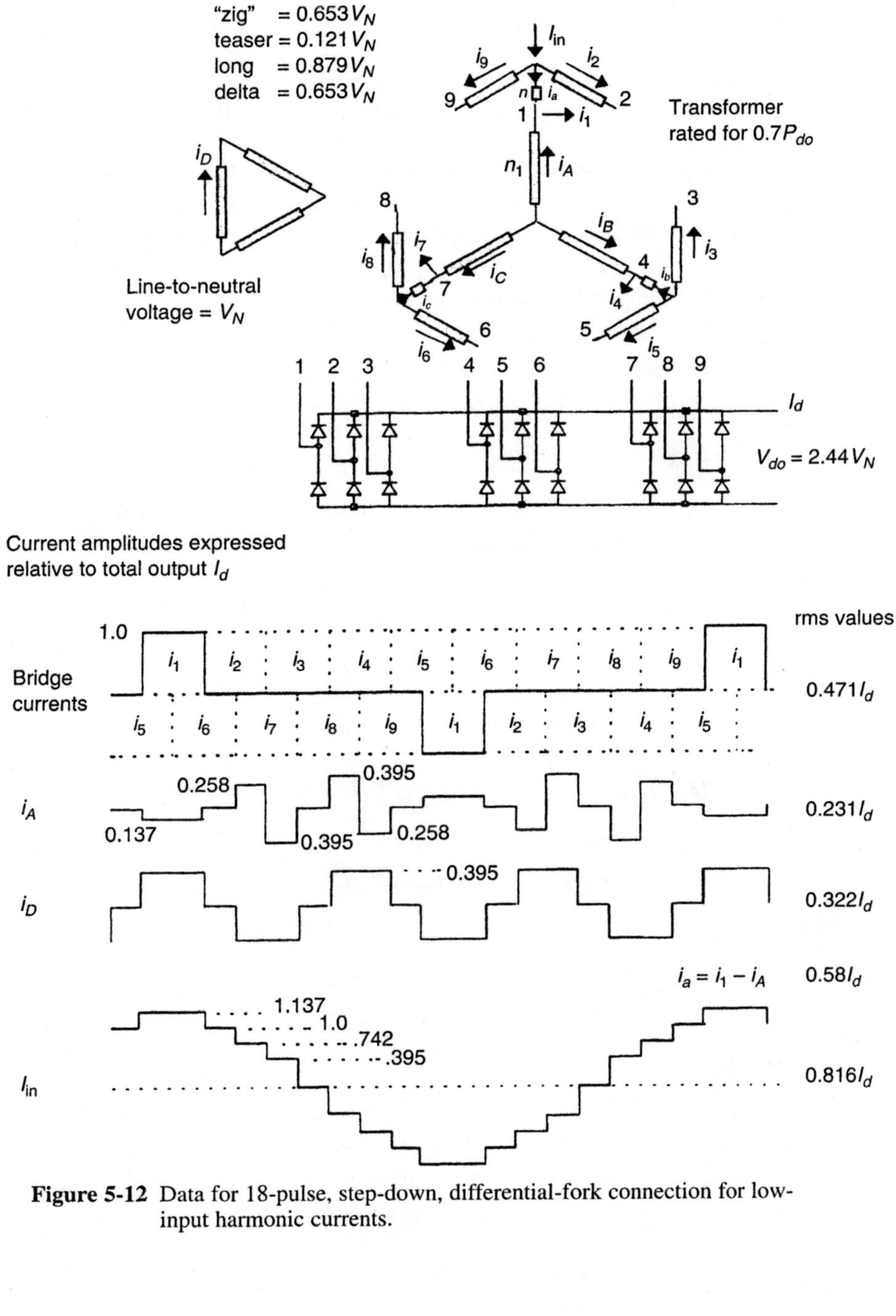

Figure 5-12 Data for 18-pulse, step-down, differential-fork connection for low-input harmonic currents.

In an idealized representation the individual output current pulses have a peak value equal to the dc output current I_d. Using equation (5.42) to determine i_A, (5.43) to calculate the delta winding current, and (5.44) to determine the input current, the temporal relationships of the currents can be defined. The resulting current patterns are drawn in Figure 5-12.

In a practical design the transformer leakage will significantly extend the device conduction period and reduce high-frequency harmonic currents. These effects are easily observed in computer simulations. This circuit has been selected for computer analysis in Chapter 9.

To determine the equivalent kVA rating the transformer voltages and currents are determined as follows:

5.5.5 Transformer Currents

From the drawings in Figure 5-12,

$$\text{current in zig winding} = 0.471 I_d \tag{5.45}$$

$$\text{current in phase winding} = 0.231 I_d \tag{5.46}$$

$$\text{current in teaser winding} = 0.58 I_d \tag{5.47}$$

$$\text{current in delta winding} = 0.322 I_d \tag{5.48}$$

5.5.6 Transformer Equivalent VA Rating

$$\text{total VA in "zig" windings} = 6 \times 0.6527 \times 0.41 V_{do} \times 0.471 I_d = 0.76 V_{do} I_d$$

$$\text{total VA in phase windings} = 3 \times 0.8794 \times 0.41 V_{do} \times 0.231 I_d = 0.249 V_{do} I_d$$

$$\text{total VA in teaser windings} = 3 \times 0.1206 \times 0.41 V_{do} \times 0.58 I_d = 0.086 V_{do} I_d$$

$$\text{total VA in delta windings} = 3 \times 0.6527 \times 0.41 V_{do} \times 0.322 I_d = 0.258 V_{do} I_d$$

$$\text{transformer total equivalent rating} = 0.68 V_{do} I_d \tag{5.49}$$

5.5.7 Analysis of Step-Down Fork for 12-Pulse Converters

A modified step-down, differential-fork connection can be used to provide two sets of six-phase voltages suitable for a superior 12-pulse connection. Figure 5-13 shows this new 12-pulse topology. It gives excellent all-around performance and tolerates the effects of type 2 power. Shown as an autoconnection in Figure 5-13, it can also be used as a double-wound configuration if power is applied to the delta windings.

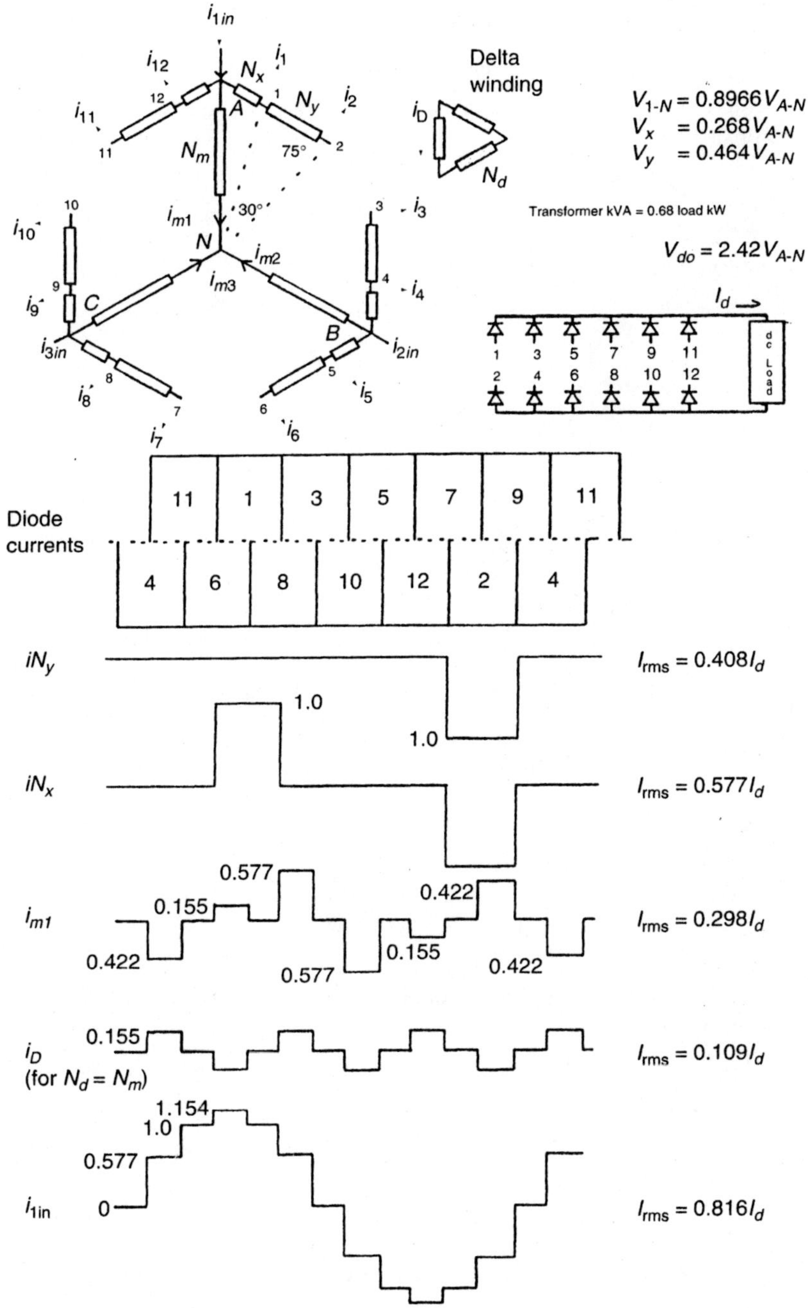

Figure 5-13 Differential fork for 12-pulse connection.

5.5.8 Analysis of Winding Voltages

For analysis refer to Figure 5-13. The winding turns are defined as N_m, N_x, N_y, and N_d. Voltages across these turns are defined as V_m, V_x, and so on. A unique combination of turns gives the required phase shift for two six-phase supplies with six windings per phase. The output voltages are at a reduced amplitude which conveniently compensates for the higher output inherent in a 12-pulse midpoint method.

Assume the output voltages such as $V_{1\text{-}N}$, $V_{2\text{-}N}$, and so on have an equal amplitude of V_o, then

$$\frac{V_o}{\sin 60°} = \frac{V_{A-N}}{\sin 105°}$$

Thus,

$$V_o = 0.8966V_{A\text{-}N} \tag{5.50}$$

$$\frac{V_y}{\sin 30°} = \frac{V_o}{\sin 75°}$$

Thus,

$$V_y = 0.4641V_{AN} \tag{5.51}$$

$$\frac{V_x}{\sin 15°} = \frac{V_o}{\sin 60°}$$

Thus,

$$V_x = 0.2679V_{A\text{-}N} \tag{5.52}$$

Also,

$$V_m = V_{A\text{-}N}$$

The delta winding circulates triplen harmonic currents, and its turns (Nd) can be any convenient number.

The dc output voltage in terms of the line-to-neutral voltages V_o and $V_{A\text{-}N}$ is given by

$$V_{\text{do}} = 2\sqrt{2}\ V_o \frac{3}{\pi} \int_{-\pi/6}^{\pi/6} \cos \omega t \, d\omega t \tag{5.53}$$

from which

$$V_{\text{do}} = 2.7\ V_o = 2.42\ V_{A-N} \tag{5.54}$$

5.5.9 Analysis of Winding Currents

For this we use Kirchoff's law of current, and the fact that the ampere-turns sum to zero. Currents flowing from the nodes define winding currents. For example, i_1 flows from node 1, i_2 from node 2, i_3 from node 3, etc. Current i_{m1} flows in phase A winding, i_{m2} in phase B, and so on. Phase winding N_m has per unit turns. For analysis, the delta winding is assumed to have per unit turns also. In practice, delta windings can have any convenient turns.

$$i_{1in} = i_{m1} + (i_1 + i_2) + (i_{11} + i_{12}) \tag{5.55}$$

$$i_{m1} - (i_3 + i_4)N_x - i_3N_y - (i_9 + i_{10})N_x - i_{10}N_y + i_D = 0 \tag{5.56}$$

$$i_{2in} = i_{m2} + (i_3 + i_4 + i_5 + i_6) \tag{5.57}$$

$$i_{m2} - (i_1 + i_2)N_x - i_2N_y - (i_7 + i_8)N_x - i_7N_y + i_D = 0 \tag{5.58}$$

$$i_{m3} - (i_{11} + i_{12})N_x - i_{11}N_y - (i_5 + i_6)N_x - i_6N_y + i_D = 0 \tag{5.59}$$

Adding (5.58) and (5.59) and using $i_{m1} + i_{m2} + i_{m3} = 0$, we obtain

$$-i_{m1} - (i_1 + i_2 + i_{11} + i_{12})N_x - i_2N_y - i_{11}N_y - (i_7 + i_8 + i_5 + i_6)N_x - (i_7 + i_6)N_y + 2_{iD} = 0$$

Thus,

$$2iN_d = i_{m1} + (i_1 + i_2 + i_5 + i_6 + i_7 + i_8 + i_{11} + i_{12})N_x + (i_2 + i_6 + i_7 + i_{11})N_y \tag{5.60}$$

Rearranging (5.56),

$$i_{m1} - N_x(i_3 + i_4 + i_9 + i_{10}) - N_y(i_3 + i_{10}) + i_D = 0 \tag{5.61}$$

Multiplying (5.61) by two and adding (5.60), we obtain

$$i_{m1} = -\left(\frac{N_x}{3}\right)(i_1+i_2-2i_3-2i_4+i_5+i_6+i_7+i_8-2i_9-2i_{10}+i_{11}+i_{12}) - \left(\frac{N_y}{3}\right)(i_2-2i_3+i_6+i_7-2i_{10}+i_{11}\cdots) \tag{5.62}$$

Putting (5.62) in (5.55), we obtain

$$
\begin{aligned}
i_{1in} = {} & i_1\left(1-\frac{N_x}{3}\right)+i_2\left(1-\frac{N_x+N_y}{3}\right)+i_3\left[2\frac{N_x+N_y}{3}\right] \\
& +i_4\left(2\frac{N_x}{3}\right)-i_5\left(\frac{N_x}{3}\right)-i_6\left(\frac{N_x+N_y}{3}\right)-i_7\left(\frac{N_x+N_y}{3}\right)-i_8\left(\frac{N_x}{3}\right)+i_9\left(2\frac{N_x}{3}\right) \\
& +i_{10}\left[2\frac{N_x+N_y}{3}\right]+i_{11}\left[1-\frac{N_x+N_y}{3}\right]+i_{12}\left(1-\frac{N_x}{3}\right)
\end{aligned}
\tag{5.63}
$$

Current in the transformer windings can be calculated as follows: main winding i_{m1} from equation (5.62), delta winding i_D from equation (5.60), and input current i_{1in} from equation (5.63). Using the turns relationships defined by (5.50), (5.51), and (5.52), we obtain

$$
\frac{N_x}{3} = 0.0893; \quad \frac{N_y}{3} = 0.1547; \quad \frac{N_x+N_y}{3} = 0.244;
$$

Results are plotted in the schematic drawing in Figure 5-13.

5.5.10 Transformer Equivalent VA Rating

$$
\begin{aligned}
\text{total VA in turns } N_m &= 3 \times 0.413V_{do} \times 0.298I_d = 0.369V_{do}I_d \\
\text{total VA in turns } N_x &= 6 \times 0.111V_{do} \times 0.577I_d = 0.384V_{do}I_d \\
\text{total VA in turns } N_y &= 6 \times 0.191V_{do} \times 0.408I_d = 0.470V_{do}I_d \\
\text{total VA in turns } N_d &= 3 \times 0.413V_{do} \times 0.109I_d = 0.135V_{do}I_d \\
\text{transformer total equivalent rating} &= 0.68V_{do}I_d
\end{aligned}
\tag{5.64}
$$

In practice the total transformer rating will be somewhat less than this due to leakage reactance effects which reduce the winding rms currents. This circuit has been defined as type MC-105 and is subjected to practical power system variations in a computer analysis given in Chapter 9. It gives excellent results.

5.6 DELTA/WYE WITH CENTER TAPPED DELTA

A combination of double-wound and autotransformers, in a hybrid connection, may be helpful in some cases. This occurs when the double-wound transformer isolation features are required for circuit topology, but the cost and losses can be reduced by the autoconnection.

Figure 5-14 shows an adaptation of a conventional delta/wye transformer that provides an effective means for obtaining voltages to feed a 12-pulse converter. The delta winding is center tapped and used as an autoconnection to feed one converter bridge at half voltage. The normal wye winding is configured to give half-output voltage that feeds another series connected converter bridge. For best results, leakage inductances should be designed to reflect as nearly as possible equal impedance to each converter.

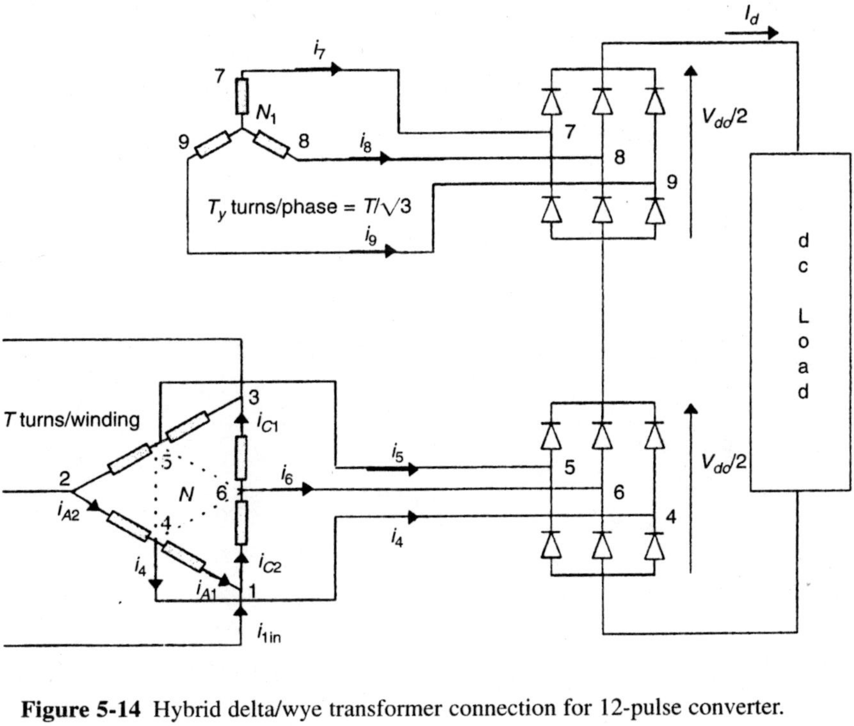

Figure 5-14 Hybrid delta/wye transformer connection for 12-pulse converter.

5.6.1 Analysis of Hybrid Delta/Wye Connection

Using the nomenclature of Figure 5-14 and setting up the equations defining zero net ampere turns, that is, $\Sigma AT = 0$; then

$$i_8 T_y + i_{A1} T + i_{A2} T = 0$$

from which

$$-i_{A1} = \frac{i_8 T_y}{T} + i_{A2} \tag{5.65}$$

From the figure,

$$i_{A2} = i_{A1} + i_4 \tag{5.66}$$

Thus,

$$-i_{A1} = \frac{i_8 T_y}{2T} + \frac{I_4}{2} \tag{5.67}$$

Also,

$$i_7 T_y + i_{C1} T + i_{C2} T = 0$$

from which

$$i_{C2} = -\frac{i_7 T_y}{T} - i_{C1} \tag{5.68}$$

From Figure 5-13,

$$i_{C2} = i_{C1} + i_6 \tag{5.69}$$

Thus,

$$i_{C2} = -\frac{i_7 T_y}{2T} + \frac{i_6}{2} \tag{5.70}$$

The supply input current is given by

$$i_{1in} + i_{A1} = i_{C2} \tag{5.71}$$

Putting (5.67) and (5.70) into (5.71) leads to the result

$$i_{1in} = \frac{-i_7 T_y}{2T} + \frac{i_6}{2} + \frac{i_8 T_y}{T} + \frac{i_4}{2} \tag{5.72}$$

For equal voltages on each converter $T_y = T/\sqrt{3}$. Thus the final result simplifies to

$$i_{1in} = \frac{i_6 + i_4}{2} + \frac{i_8 - i_7}{2\sqrt{3}} \tag{5.73}$$

To plot the current waveform represented by equation (5.73), the relative phase position of the converter current must be established. To do this, voltage vectors are drawn in the figure. From these it is determined that the current i_4 (in set i_4, i_5, i_6) leads the current i_9 (in set i_7, i_8, i_9) by 30°.

With the currents drawn in their correct phase position the waveshape for line input current can be obtained. This is shown in Figure 5-15 and is observed to be of the characteristic 12-pulse shape.

This circuit is subjected to practical power system variations in a computer analysis given in Chapter 9. The circuit is reasonably insensitive to preexisting source harmonic voltages and provides a practical 12-pulse connection. For best results the leakage inductances must be reasonably well balanced.

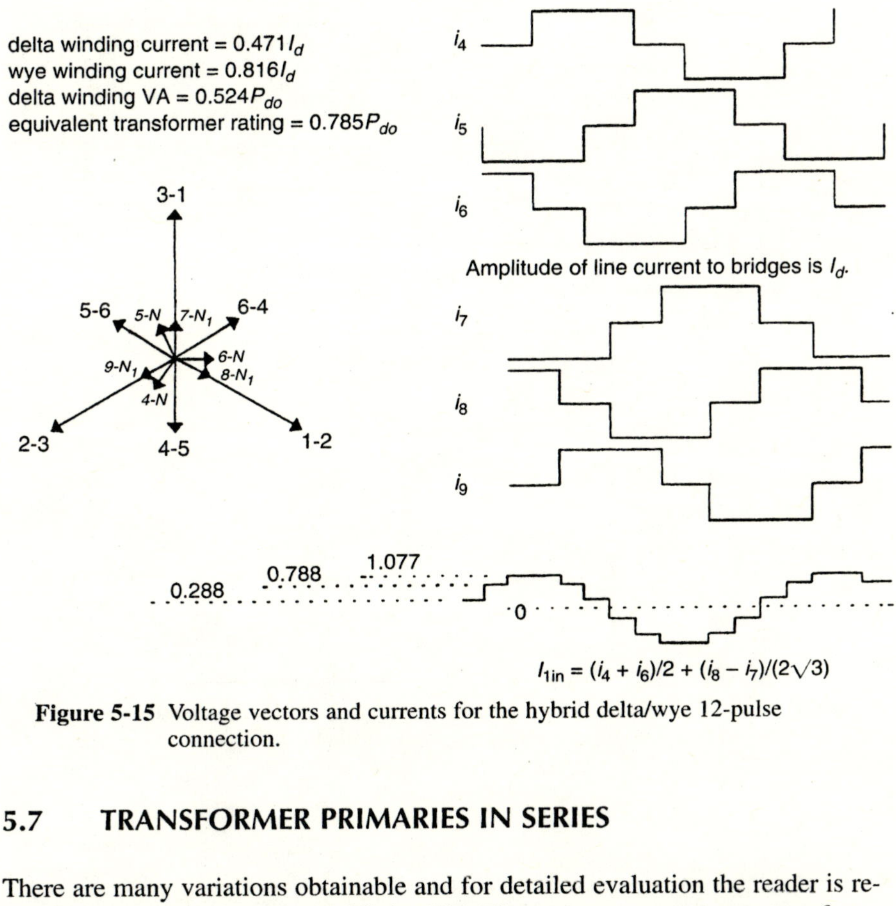

Figure 5-15 Voltage vectors and currents for the hybrid delta/wye 12-pulse connection.

5.7 TRANSFORMER PRIMARIES IN SERIES

There are many variations obtainable and for detailed evaluation the reader is referred to the work of April and Olivier [25]. The concept uses separate transformers in which the primary windings are connected in series and the secondary converters are in parallel. The secondary circuits are forced to share current by reason of the balance maintained by the transformer ampere turns. More than this, the need to balance ampere turns modifies the current harmonic content, the conduction angles of individual semiconductor devices, and transformer voltage waveforms. The resulting characteristics are similar to those obtained with harmonic blocking current transformers discussed in Chapter 6, Section 6.4. Where the required dc output is half that of a straight-through connection, one of the transformers can be omitted, as shown in Figure 5-16.

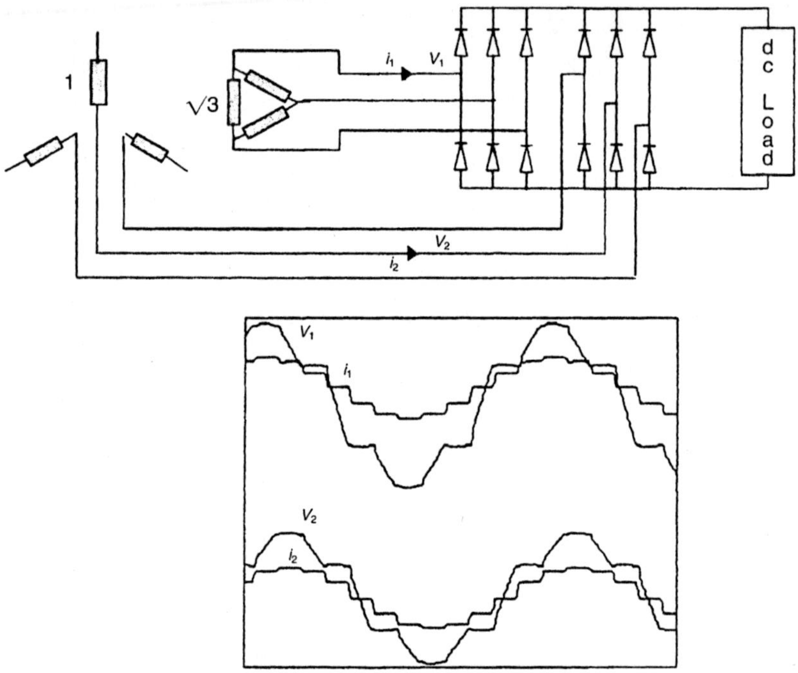

Figure 5-16 Twelve-pulse with series primary circuits and parallel converters. Transformer rating is about half power.

Chapter 6

Interphase and Current-Control Transformers

6.1 INTRODUCTION

Many of the popular multipulse converter circuits are formed by paralleling lower-pulse-number converters. For instance, 18-pulse can be obtained by paralleling three 6-pulse bridge converters.

A major design goal in multipulse operation is to get the converters, or converter semiconductor devices, to share currents equally. If this is achieved, then maximum power and minimum harmonic currents can be obtained.

To address the characteristics that facilitate successful parallel operation, we have to consider dc and ac conditions. Also, the circuit connections are important if unwanted conduction paths are possible. Figures 6-1 through 6-3 illustrate basic circuit features that affect how currents in parallel circuits are shared and help set the stage for discussion. In these figures, the power sources are effectively isolated such that there are no alternative unwanted conduction paths.

Figure 6-1 shows two batteries in parallel, feeding a common load. This figure is used to illustrate the effects of dc unbalance. For i_1 and i_2 to be equal, we need E_1 and E_2 to have the same amplitude and resistors r_1 and r_2 to be equal. Alternatively, if E_1 and E_2 are not equal, then appropriate values of r_1 and r_2 can be chosen to give equal currents under a given load condition. The voltage regulation diagram shows that for different loads the battery currents will be different.

Figure 6-2 shows two ac voltage sources in parallel. Just as resistance does in dc circuits, ac sources can have impedance to share the load currents. However, by joining the circuits through a tapped current transformer (CT), a powerful method of controlling current balance is possible. If turns n_1 and n_2 are equal, then currents i_1 and i_2 will be equal, provided that the current transformer is designed with the ability to supply the compensating voltage.

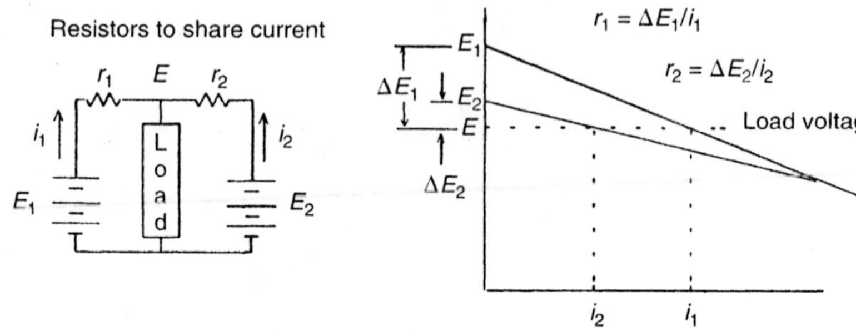

Figure 6-1 Paralleling dc voltage sources.

In Figure 6-2 the transformer plays an active role by developing a voltage to correct for any unbalance caused by source voltage or impedance variations. If the series source impedance is low, then any voltage difference $(V_1 - V_2)$ is dropped across the transformer. The load voltage in this case becomes the average of the two source voltages, namely, $(V_1 + V_2)/2$. There is no stipulation that

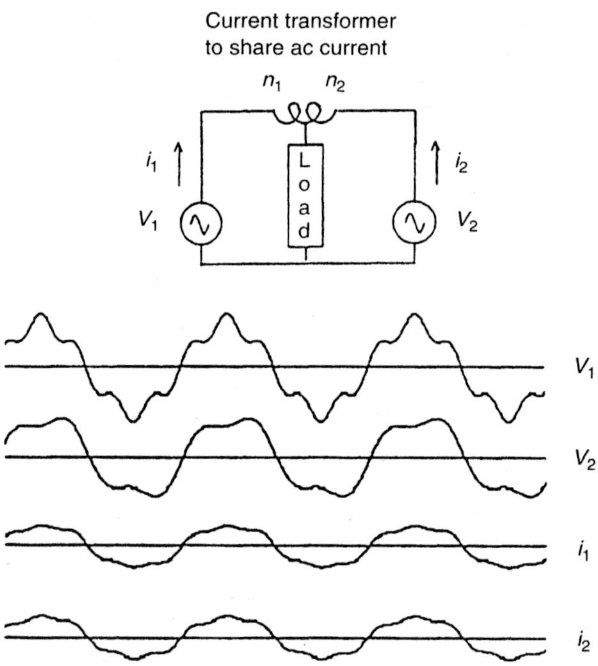

Figure 6-2 Paralleling ac voltage sources.
Waveforms for $n_1 = n_2$ and resistor load.

the voltages or currents are sinusoidal; however, they are assumed to be alternating at a frequency compatible with the ability of the transformer to supply flux changes and voltage drop.

Figure 6-3 shows two three-phase converter circuits paralleled in conjunction with an interphase transformer. In these circuits we must address both dc and ac effects.

Each converter includes an ac output voltage ripple, which is a natural part of the power conversion process. This ripple is affected by ac line reactance and (where used) by phase control. Even if the converters have identical ripple voltage patterns, they will, in the case of multipulse connections, have instantaneous

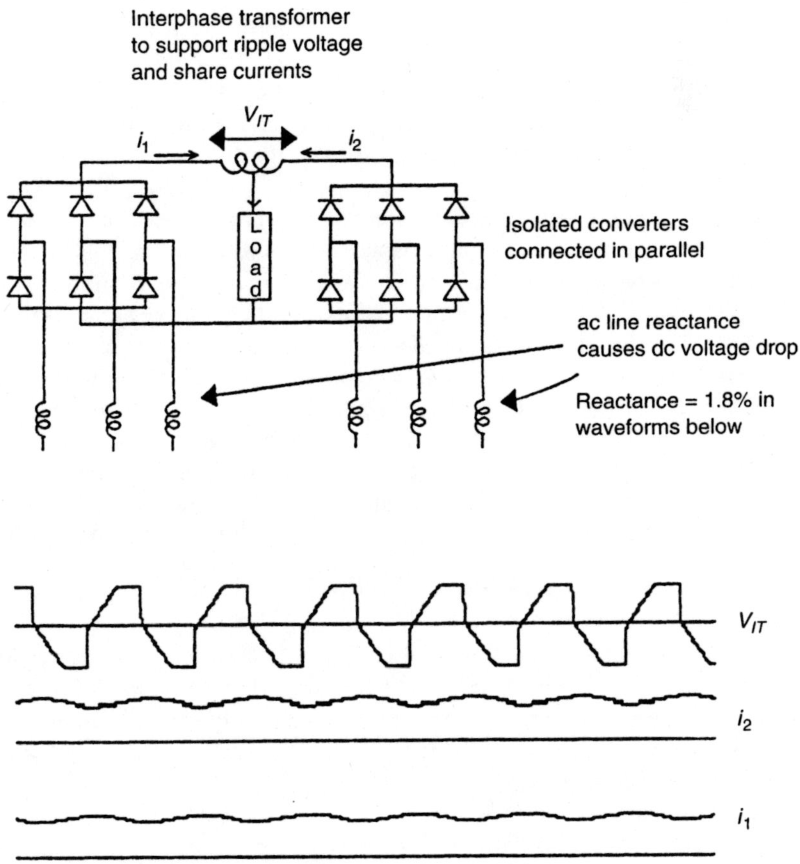

Figure 6-3 Paralleling of sources with ac and dc voltages (in waveforms, magnetizing current of interphase transformer causes 18% peak-to-peak ripple current in i_1 and i_2).

differences because of the phase shift in each source. If these voltage differences are not supported in some way, the device conduction patterns will be changed, and each converter will interfere with the operation of the other. To prevent this interfering effect, an interphase transformer is interposed between the two converters. This interphase transformer is shown in Figure 6-3. The terms interphase reactor, spanning reactor, and current-balancing transformer have also been applied to describe this device.

Reactance in each converter ac line causes a voltage drop at the dc output in each circuit, just as the resistors do in the case of paralleled batteries but without attendant power dissipation. Any resistance drop associated with each converter causes the same effect. Together these dc regulation effects provide some assistance toward current sharing.

For practical values of ac line reactance and converter resistance, the effective source resistance may be quite small. For example, the no-load to full-load dc voltage drop may be 3% or less. If the voltage drop is 3%, then assuming that each converter must not exceed its current rating, a 1% difference in converter dc voltage would result in one converter supplying 50% more current than the other.

It is highly desirable that the converters have nearly identical dc output voltages. However, it has been shown in Chapter 2, Section 2.6.2, that this is not simply a matter of converter design. System factors, such as preexisting harmonic voltages, are involved. For this reason it is good practice to have some form of dc control to achieve balance in paralleled converters.

Interphase transformers are also used in converter circuits in which the ac voltage sources are not isolated. Here, in addition to supporting ripple voltage, they also prevent unwanted conduction paths. The voltage drop across the interphase transformer(s) is usually two or three times as great in this type of connection. This results in a more difficult and larger interphase transformer design.

6.2 INTERPHASE TRANSFORMERS

The interphase transformer(s) functions to support ac voltage differences existing between converter outputs and allows the converters to act as if they are operating alone. However, they cannot help balance steady-state differences in dc voltage. They support instantaneous (i.e., ac) voltage differences but not average (i.e., dc) voltage differences.

The windings carry both dc and ac currents. Connections are made such that dc currents, flowing to the load, cause opposing ampere-turns on the core. There is no inherent restriction on the number of windings or converters that may be paralleled, provided the magnetic core has an appropriate number of limbs.

6.2.1 Combining Three Converters

Figure 6-4 shows how three bridge converters might be paralleled, as in an 18-pulse system. When the currents are balanced, there are no net dc ampere-turns (*AT*s) acting directly around the flux loops. However, the *AT*s are in the same sense in each leg and act through the high-reluctance air path outside the core to produce a small amount of leakage flux. Unbalanced currents can cause substantial dc flux. The total flux must not be so high as to cause saturation of the iron core.

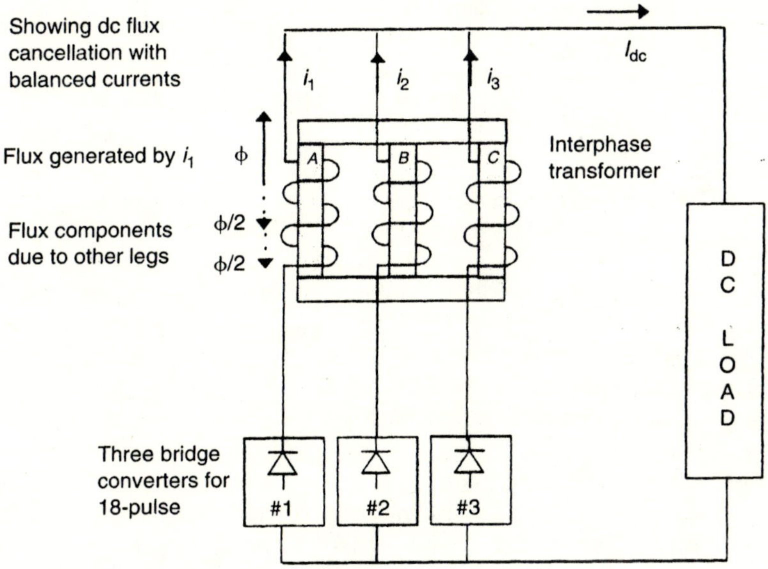

Figure 6-4 Interphase transformer for three converters.

6.2.2 Combining Two Converters

A two-winding interphase transformer is included in the 12-pulse converter arrangement in Figure 6-5. This transformer can be constructed as in Figure 6-4 but with only two limbs. Alternatively, a single-phase shell structure can be used.

The advantage of a shell mechanical construction (shown in Figure 6-10) is that the windings on the interphase transformer can be trifilar wound and placed on the center limb. This gives excellent coupling. When currents are balanced and appropriate start/finish connections are made, the dc ampere-turns are effectively balanced at the source. Because of this, negligible leakage flux returns outside the core.

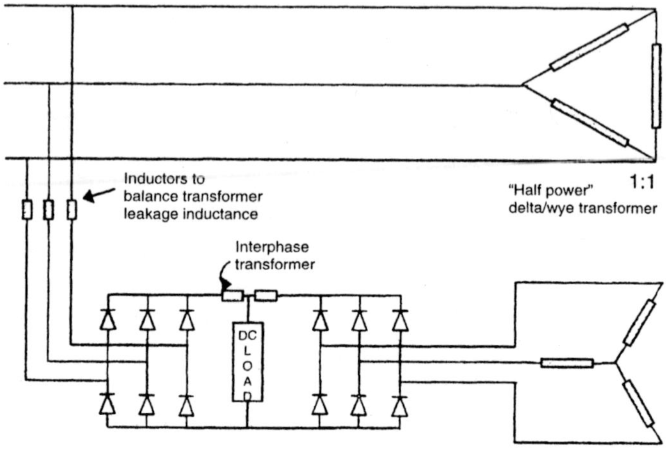

Figure 6-5 Showing an interphase transformer to support ripple between two paralleled converters.

The ac ripple voltage existing between the converters can be calculated from the specific converter output waveforms, but, as noted earlier, the effects of commutating reactance and any phase control must be considered. Figure 6-6 shows these calculated values for two 30°-displaced 6-pulse systems [1].

The value of E_x in this figure refers to the reduction of average value of direct voltage caused by commutation.

Volt-seconds in the ripple voltage will produce flux swings and a corresponding magnetizing current to flow in the interphase transformer. This magnetizing current contributes to the ripple current flowing in each converter such that even if the total load current is constant the converter output currents are not. This effect is clearly seen in Figure 6-3. The ripple current modifies the harmonic currents in each converter.

For a given connection, the magnetizing current depends upon the interphase transformer magnetizing inductance. The design is chosen so that the worst case magnetizing current does not interfere with current sharing at the lowest required load. When the dc load current is reduced to an average value that is less than the transformer peak magnetizing current, the mode of operation changes. The performance under these low output conditions is affected by the core magnetic characteristics and tends to be rather unpredictable.

For the circuit in Figure 6-5, satisfactory current sharing is feasible over more than 5:1 range of load if the peak-to-peak magnetizing current (of total transformer turns) does not exceed 20% of maximum dc load current.

An airgap may be used in the magnetic path of the interphase transformer to help prevent the core from saturating under dc current unbalance conditions. However, this causes an undesirable effect in that it reduces the magnetizing in-

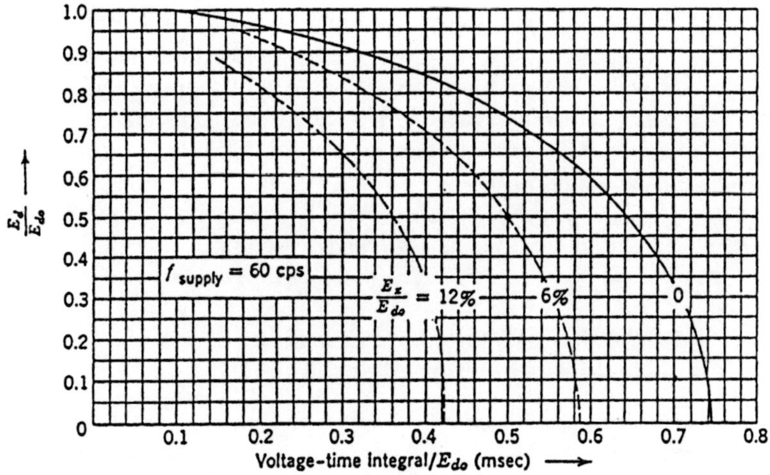

Figure 6-6 Time integral of voltage absorbed by interphase transformer if combining two 30°-displaced 6-pulse systems. (Source: Schaeffer, *Rectifier Circuits: Theory and Design.* Copyright © 1965 Wiley-Interscience. Reprinted by permission of John Wiley and Sons, Inc.)

ductance. The final interphase transformer design involves a number of equipment design trade-offs. As noted in Section 2.6, the final performance may be adversely affected by system parameters. Many practical features need to be considered.

The 12-pulse arrangement in Figure 6-5 was first introduced in Figure 2-6. It is worthy of additional comments here. This arrangement is practical only when the required dc output voltage is the same as that given by the straight-through connection. The transformer provides isolation and a 30° phase shift for the 12-pulse performance. Because of the isolation, the ripple voltage supported across the interphase transformer is significantly lower than in some nonisolated circuits.

If the converter currents are evenly shared, the transformer need be rated only for 50% of the effective kVA supported; however, without means to ensure proper current sharing, the rating may easily be 75% of the power. There is discussion of variations to this circuit later on in this chapter.

6.2.3 Effects of Interphase Transformer Saturation

In general, any saturation of the interphase transformer will detract from the design performance; however, this is more critical in some circuits than others. In Figure 6-5, for example, the interphase transformer is not essential to secure the 12-pulse action. Rather, it is a means to secure the desirable features of each converter acting alone, for example, a 120° device conduction.

Without an interphase transformer in Figure 6-5, or if it is totally saturated, the rectifier conduction angle will, neglecting ac line reactance, change from that of a single 120° pulse to two 30° pulses. In some cases, this change in basic operating mode can be tolerated, especially if the circuit ac reactance is sufficient to significantly extend the device conduction at full load. Transformer reactance of 5% is adequate to achieve this.

In some multipulse connections, especially when the 6-pulse constituents are not isolated, the interphase transformer is essential for 12-pulse operation. This is a disadvantage and requires even more careful control of the design.

6.2.4 Paralleling When the Phase Shift Is by Means of an Autotransformer

If the 30° phase shift, appropriate to 12-pulse performance, is obtained using a polygon transformer, as in Figure 5-1, the transformer rating will be only half that given in equation (5.11), namely, 22% of the total dc output power. A dual output polygon transformer is rated for about 19%. Figure 6-7 shows an exam-

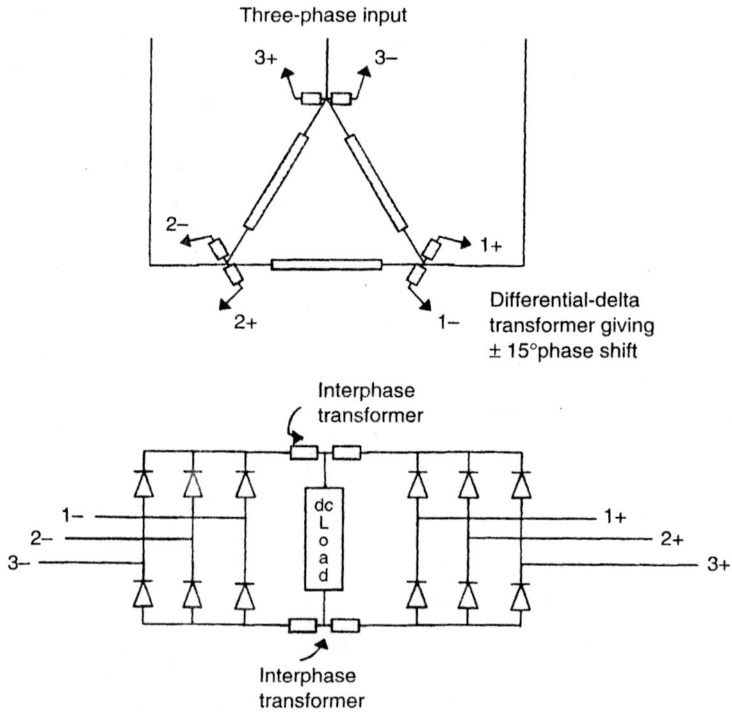

Figure 6-7 Two interphase transformers allow 12-pulse with auto-connected phase-splitting transformer.

ple of this topology in which a differential delta transformer is used to provide the phase shift. The principle applies to any autotransformer.

To obtain 12-pulse results from the circuit in Figure 6-7, two interphase transformers are essential. One interphase transformer is connected in the positive output, and one is connected in the negative output. These interphase transformers isolate unwanted conduction paths and allow the converters to operate practically independently.

Because the autotransformer does not provide isolation, the voltage to be supported across the interphase transformer is much greater. Each interphase transformer is larger than the single unit in Figure 6-5.

The basic device conduction is 120°, and the total ac line input current is that of a conventional 12-pulse converter. The small size of the phase-shifting transformer makes this an efficient arrangement. Designs of this type have given good practical service.

If the interphase transformers are not incorporated in Figure 6-7 or if significant saturation occurs, the circuit regresses to give undesirable 6-pulse operation. Figure 6-8 shows the connections, waveforms, and voltage vectors relevant to this condition.

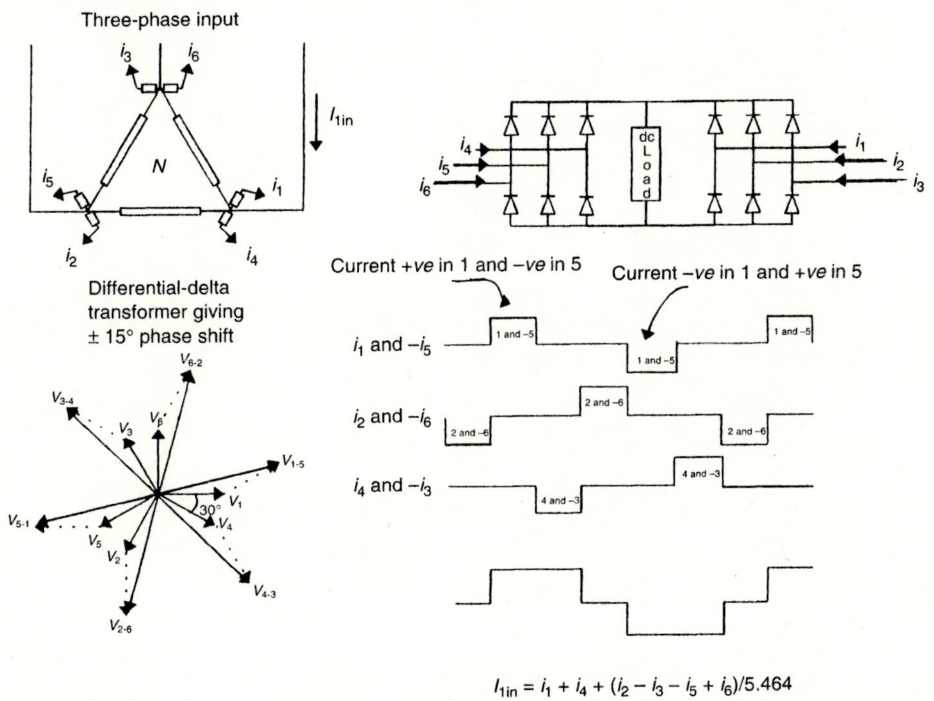

Figure 6-8 Illustrating interaction of converters without interphase transformers.

Referring to these figures, there are observed to be unwanted paths such that the 12 devices of the two converter bridges are caused to conduct in a sequence that is different from a conventional 12-pulse arrangement. This occurs due to higher than normal line-to-line voltages developed across certain of the bridge terminals. For example, voltages such as $V_{1\text{-}5}$ existing between the bridges are about 11% greater than the normal line voltages, such as $V_{1\text{-}3}$, applied to the bridges. These voltages force the unfavorable conduction paths.

With this mode of operation the devices conduct for a period of 60°, and the normal 12-pulse operation is not achieved. The eventual ac line current is determined to be of 6-pulse shape. From this understanding, a modified approach to the interphase transformer design becomes apparent in which a separate transformer is used to suppress the unwanted conduction paths. This is discussed in Section 6.3.1.

Two basic methods can be considered for applying autotransformers to achieve multipulse effects. They can be used to parallel discrete 6-pulse converter circuits, in which case interphase transformers are required. Alternatively, they can be applied in a manner that reduces the semiconductor conduction intervals, such as in midpoint connections. In the latter case, interphase transformers are unnecessary because there are no alternative conduction paths.

Application of reduced-rating autotransformers to midpoint connections has been used very successfully. For example, Figure 6-9 shows a phase-splitting

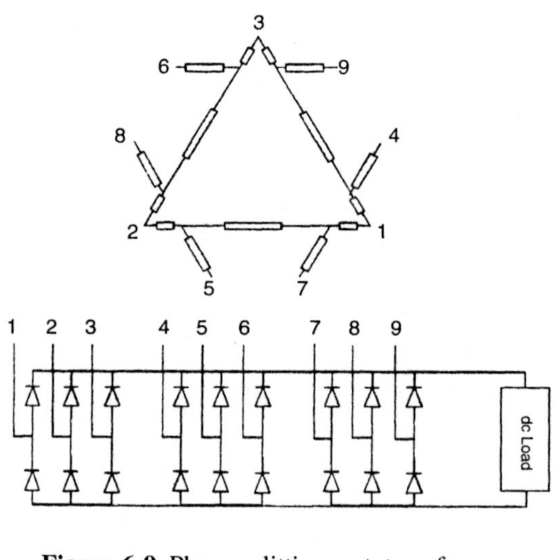

Figure 6-9 Phase-splitting, autotransformer connection for multipulse operation without interphase transformers.

autotransformer connection applied to give 18-pulse operation in a nine-phase bridge arrangement.

6.3 CURRENT-BALANCING TRANSFORMERS

The interphase transformers shown so far have been connected in the dc output of the converters. Their performance is not crucial when the converter power sources are isolated but becomes vital when multiple bridges are paralleled in conjunction with auto-connected phase-shifting transformers. In this latter case an alternative technique can be applied in which a three-winding current transformer is used to provide comparable isolation.

6.3.1 Zero-Sequence Effects

As was seen in Figure 6-8, the autotransformers disable 12-pulse action because they allow currents to flow without hindrance from one bridge to the other. These currents contain significant amounts of 3rd harmonic current. In contrast, use of a delta/wye double-wound transformer in Figure 6-5 automatically suppresses flow of these 3rd harmonic currents. Thus, it is reasoned that equivalent 12-pulse performance can be approximated with phase-shifting autotransformers, if a means is provided for suppressing 3rd harmonics of current.

In a balanced three-phase system, the 3rd and other triplen harmonics of current or voltage are in phase with each other in each line. (Triplen derives from triple "n" and includes odd multiples of three, such as 3, 9, 15.) For example, a 3rd harmonic in line *A* will have a counterpart in line B at an angle of $3 \times 120°$, that is, 360°. Thus, the 3rd harmonics appear as a zero-sequence set at three times frequency. Certain transformer connections, such as in the zero-sequence isolator, have the ability to suppress zero-sequence currents, especially when they are wound with a shell structure that provides a low-reluctance path for the magnetic flux [20]. We can use one of these connections to effectively eliminate, or at least significantly reduce, the 3rd harmonics of current that compromise the 12-pulse characteristics of paralleled converters fed from autotransformers.

The converter line currents represent a three-phase set with fundamental currents displaced by 120°. These sum to zero. When the 3rd harmonics are balanced, they are all in phase and can sum to zero only if each is zero. Thus, a current transformer with three isolated coils that sum the three currents will pass positive- and negative-sequence currents but suppress 3rd and other triplen harmonics.

This transformer is effectively described as a zero-sequence blocking transformer (ZSBT) because it exhibits high impedance to zero-sequence currents. As noted previously, it is most effective in a shell-type structure that provides a low reluctance path for all fluxes. A core form of construction will also provide suppression of zero-sequence components but since the 3rd harmonics are in the

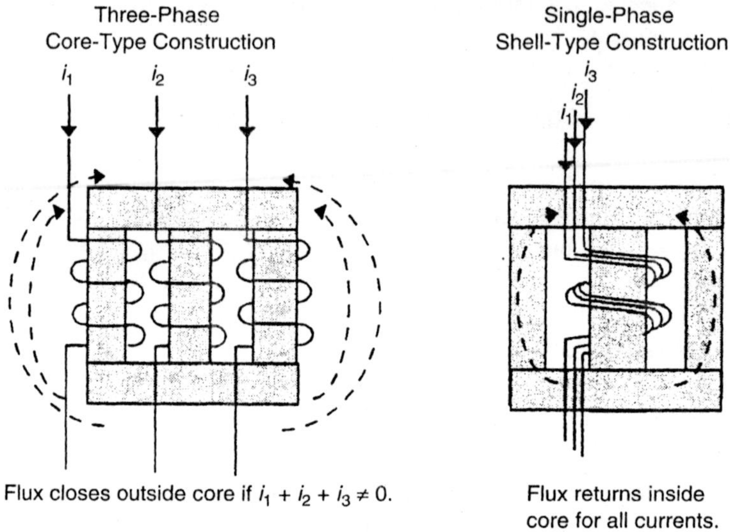

Figure 6-10 Showing effects of transformer construction on zero-sequence fluxes.

same phase, their flux must return outside the core. Because of this, the transformer zero-sequence impedance is reduced. These two examples are illustrated in Figure 6-10.

The ZSBT represents the first step in deriving a series of harmonic current blocking transformers and is shown in Figure 6-11 applied to the 12-pulse system. In this first application example, the circuit functions just like the circuit of Figure 6-5.

Without an interphase transformer in the dc outputs, the converter device conduction angle is 60°, composed of two 30° pulses. However, as pointed out earlier, ac line reactance usually increases this conduction period significantly. A single interphase transformer can also be used to extend the current conduction angle, just as in Figure 6-5. This is shown in connection MC-102A in Chapter 9.

The simplified schematic of Figure 6-11 shows only one turn per phase, whereas in practice there may be several turns per phase. The voltage dropped across the core is almost entirely 3rd harmonic.

The equivalent double-wound transformer rating for this zero-sequence blocking transformer depends upon a number of factors. However, its design is straightforward, and it has a power rating which is usually less than 3% of the dc output power.

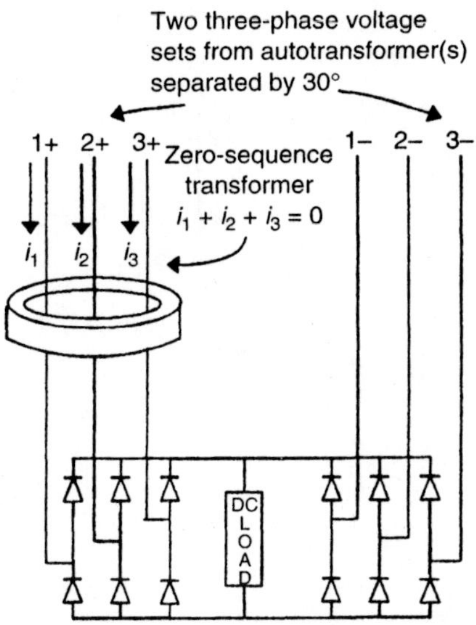

Figure 6-11 Zero-sequence blocking transformer to isolate parallel bridges fed by autotransformer.

6.3.2 ZSBT for 18-Pulse Operation

For 18-pulse operation using auto-connected phase-shifting transformers, at least two ZSBTs are required to block all possible 3rd harmonic conduction paths in the converter circuits. One possible schematic is shown in Figure 6-12.

In this 18-pulse circuit the polygon transformers provide a phase shift of 20°. Each has a rating equal to only 11% of the dc load power. The zero-sequence blocking transformers have a rating of less than 2.5% of the dc load power. The "natural" device conduction period, without an interphase transformer in the dc output, is 40° consisting of two 20° pulses. Practical values of ac line reactance, for example, 1.8%, extend this conduction to 120° at full load.

Like other multipulse circuits that operate by paralleling multiple converter bridges, these circuits are affected by preexisting line harmonic voltages. Computer-calculated results for the case of a preexisting 5th harmonic of voltage, at zero phase angle, are shown in Figure 6-13.

As discussed previously, the effects of preexisting harmonic voltages can be mitigated by regulating the converter outputs such that each bridge converter draws the same amount of current.

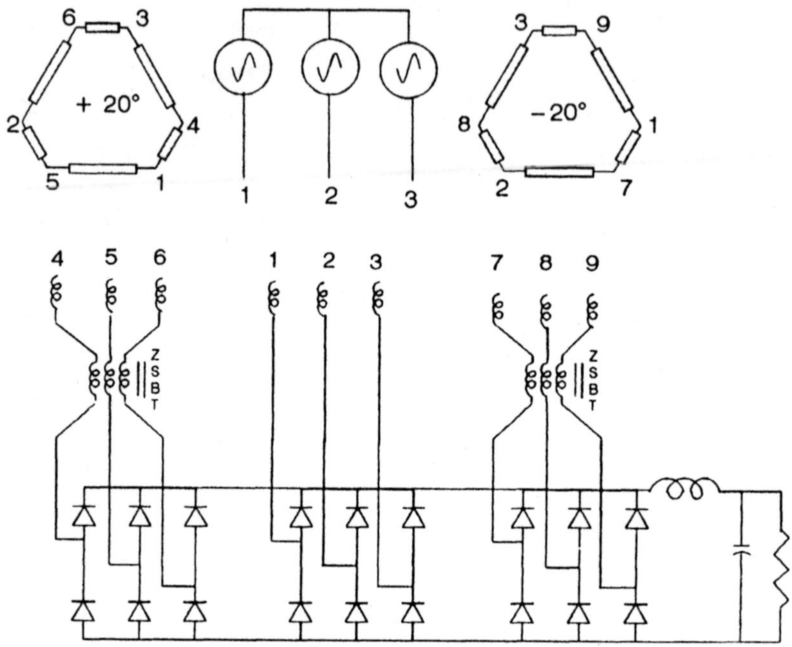

Figure 6-12 Eighteen-pulse converter with polygons and ZSBTs.

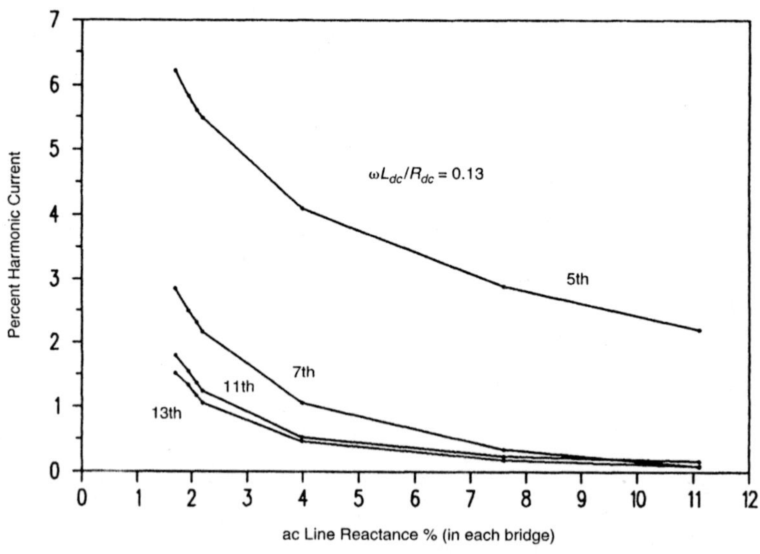

Figure 6-13 The effects of 2.5% preexisting 5th harmonic of voltage in the source for the circuit in Figure 6-12.

6.4 HARMONIC BLOCKING CURRENT TRANSFORMERS

Harmonic blocking transformers are extensions of the zero-sequence blocking transformer. By cross-connection of appropriate line currents, they block certain positive and negative-sequence harmonics and triplen harmonics. In doing so, they automatically extend the conducting period of the converter devices.

The technique of using current transformers to sum various currents has been used in feeder protection schemes [21] and in voltage regulating systems for the aircraft industry. The current potential transformer (CPT) is an example applied to generator exciters. This transformer sums alternator line currents in conjunction with a quadrature current to produce a resultant current, which is rectified and fed to the field system. The summing automatically produces excitation sensitive to the load power factor.

Current transformers for harmonic current cancellation appear to have been first proposed by Udo H. Meier in his 1971 U.S. Patent #3,792,286.

The circuit in Figure 6-2 shows a current transformer employed to parallel two single-phase circuits. In this case only two windings are required, and the connection is straightforward. For current transformers to help share the currents in two three-phase circuits operating at different phase angles, more windings are required. Specifically, the current transformers have windings that monitor three currents that must sum to zero under the required conditions.

In Meier's patent, the focus was on means for sharing the output current of multiple variable-frequency inverters; however, the principle applies in exactly the same way for input converters. Current transformers used to cross connect two three-phase converter systems to produce 12-pulse are shown in Figure 6-14.

To determine the manner of operation of harmonic blocking current transformers, the vector diagram shown in Figure 6-15 is used. This represents one method of connection for two three-phase systems. The windings are arranged such that, when the fundamental currents sum to zero, the desired balance conditions are obtained. Some of the harmonic currents do not sum to zero and are therefore "blocked" by the CT impedance.

To combine two three-phase bridge converters, the currents i_1, i_2, and i_3 flow in one converter, and currents i_4, i_5, and i_6 flow in the other. The turns ratios can be chosen to provide the required harmonic blocking affects. For 12-pulse operation the current transformers have three sets of turns with ratios of $\sqrt{3}$:1:1. The winding with $\sqrt{3}$ turns carries current from one three-phase set of currents. Two equal-turn windings on the same core are influenced by two currents from the other three-phase set, as shown in Figure 6-14.

The vector drawing shows that, by using i_1 in conjunction with a $(1/\sqrt{3})$ fraction of the currents i_4 and i_6, the fundamental currents sum to zero. Thus, there is negligible impedance presented to the fundamental.

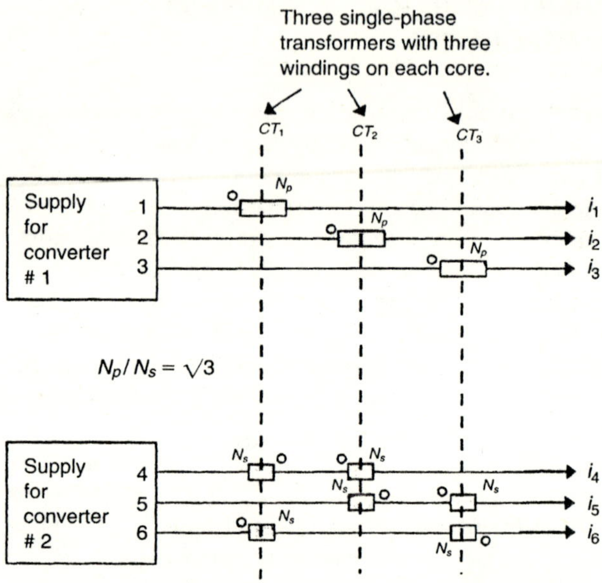

Figure 6-14 Harmonic blocking current transformers interconnecting two three-phase systems.

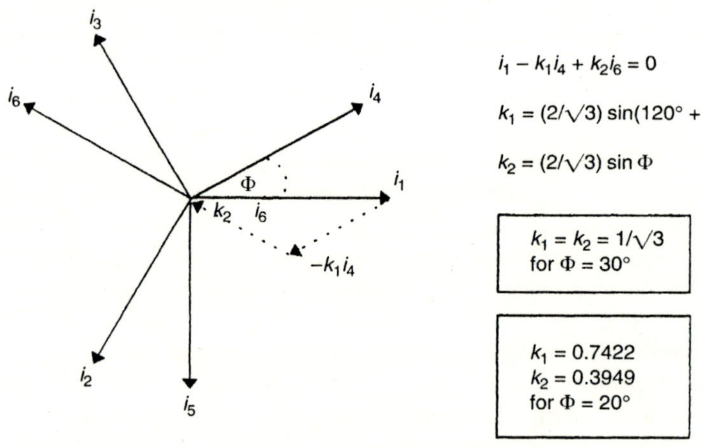

Figure 6-15 Vector drawing showing cancellation of fundamental ampere-turns.

In Figure 6-14, the phase angles of the supply voltage for converter #1 and converter #2 are not defined. From the vector drawing in Figure 6-15, the inference is that the voltages fed to converter #2 will lead those fed to converter #1 by an angle of Φ, with Φ being 30° for 12-pulse. Although reducing the fundamental voltage dropped across the CT may be desirable, it is not essential. To satisfy the CT ampere-turn balance, appropriate voltages are automatically dropped across the current transformer. The performance of the CTs with respect to harmonic frequencies is the means whereby "blocking" of the harmonic currents is achieved.

To determine this performance, the vector position of the harmonic currents must be defined. These are given in Table 6-1 for a 12-pulse arrangement. The current i_1 and its harmonics are shown in the reference position of $(1 + j0)$. If the fundamental is at an angle of Φ, then the nth harmonic will be at an angle of $n\Phi$.

TABLE 6-1

CURRENT VECTORS IN 12-PULSE CONVERTER					
	1st	**3rd**	**5th**	**7th**	**11th**
i_1	$1 \angle 0$	$1 \angle 0$	$1 \angle 0$	$1 \angle 0$	$1 \angle 0$
i_2	$1 \angle -120$	$1 \angle 0$	$1 \angle 120$	$1 \angle -120$	$1 \angle 120$
i_3	$1 \angle 120$	$1 \angle 0$	$1 \angle -120$	$1 \angle 120$	$1 \angle -120$
i_4	$1 \angle 30$	$1 \angle 90$	$1 \angle 150$	$1 \angle -150$	$1 \angle -30$
i_5	$1 \angle -90$	$1 \angle 90$	$1 \angle -90$	$1 \angle 90$	$1 \angle 90$
i_6	$1 \angle 150$	$1 \angle 90$	$1 \angle 30$	$1 \angle -30$	$1 \angle -150$

To illustrate the harmonic current cancellation process, consider current transformer CT_1 in Figure 6-14. The net ampere turns acting on its core are given by

$$\text{core ampere-turns} = \sqrt{3}\, i_1 - i_4 + i_6$$

These ampere-turns can be calculated using the appropriate vectors from Table 6-1, after they are converted to the form $A + jB$, where $A = \cos \Phi$ and $B = \sin \Phi$. The calculations are summarized in Table 6-2.

Table 6-2 shows that 3rd, 5th, and 7th harmonics of current are "blocked" by the current transformer because their ampere-turns do not sum to zero. On the other hand, the fundamental and 11th harmonics are passed through without impedance. These calculations demonstrate the principle whereby different harmonics are blocked.

For the given example which blocks harmonics of the form $6(2k - 1) \pm 1$, the resultant current that passes is of a 12-pulse type. Thus, each converter bridge now draws 12-pulse type currents. The conduction pattern in the converter devices is different; also, because some of the harmonics are blocked, there has been a small reduction in the dc output voltage.

TABLE 6-2.

SUMMATION OF AMPERE-TURNS ON HARMONIC BLOCKING CT_1				
Harmonic	$\sqrt{3}\, i_1$	$-i_4$	i_6	ΣAT
1st (fundamental)	$\sqrt{3} + j0$	$-(\sqrt{3}/2) - j/2$	$-(\sqrt{3}/2) + j/2$	0 (passes)
3rd	$\sqrt{3}$	-1	1	$\sqrt{3}$ (blocks)
5th	$\sqrt{3} + j0$	$(\sqrt{3}/2) - j/2$	$(\sqrt{3}/2) + j/2$	$2\sqrt{3}$ (blocks)
7th	$\sqrt{3} + j0$	$(\sqrt{3}/2) + j/2$	$(\sqrt{3}/2) - j/2$	$2\sqrt{3}$ (blocks)
11th	$\sqrt{3}$	$-(\sqrt{3}/2) + j/2$	$-(\sqrt{3}/2) - j/2$	0 (passes)

The current transformer technique can be applied to help share currents in parallel converter circuits. For example, when applied as in Figure 6-16, it acts to balance the fundamental currents in the two converters. In this conceptual application, the harmonic blocking effects are not the main purpose of the current transformers. A 12-pulse input of line current is obtained by combining the straight-through converter with a phase-shifted converter. By using CTs, this 12-pulse circuit is made less sensitive to preexisting source voltage harmonics and other factors that might cause unbalance.

I am not aware of this circuit being used in practice. It performed well in a computer simulation where the zero-sequence impedance was easily defined. It is given here to demonstrate the flexibility of the current transformer balancing/harmonic blocking principles.

Waveforms were obtained from a computer simulation in which the dc load current was kept practically ripple-free by means of dc circuit inductance.

The connection in Figure 6-16 also takes advantage of the delta/wye transformer leakage reactance. The current transformers effectively make the "straight-through" converter a slave to the load on the delta/wye transformer.

In the preceding example, the current transformers were shown to force a 12-pulse pattern of currents into each three-phase converter bridge, whether it was required or not. Similar results could be expected by applying CTs to the 18-pulse circuit shown in Figure 6-12. In this case, the CT ratios would be selected to cancel fundamental currents at 20° using the equations defined in Figure 6-15.

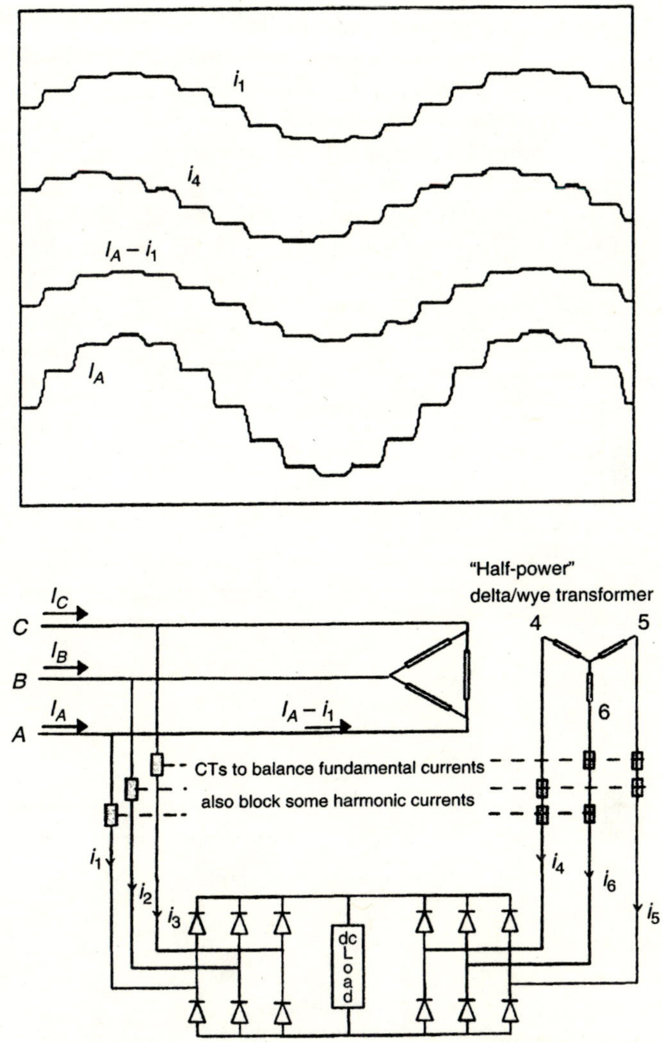

Figure 6-16 Conceptual application of current transformer.

One method of connection would use CTs to force the 0° converter and +20° inverter to balance. Another set of CTs would balance the 0° and −20° current sets. There are many variations on this scheme. One variation on this technique includes the use of zero-sequence CTs to facilitate construction of the harmonic blocking transformers using three-phase, three-limb cores.

The phase-shifting transformer is not an essential part of some of these multipulse arrangements. Figure 6-17 shows a basic connection for 18-pulse characteristics without using a phase-shifting transformer. Results from a computer simulation of this arrangement are shown in Figure 6-18. The 18-pulse waveform is seen to be developed in the individual converter ac lines as well as in the total input. The use of ZSBTs simplifies the current transformer design.

I am unaware of any tests that would confirm the practical viability of this particular approach and this connection may not be as cost effective as the

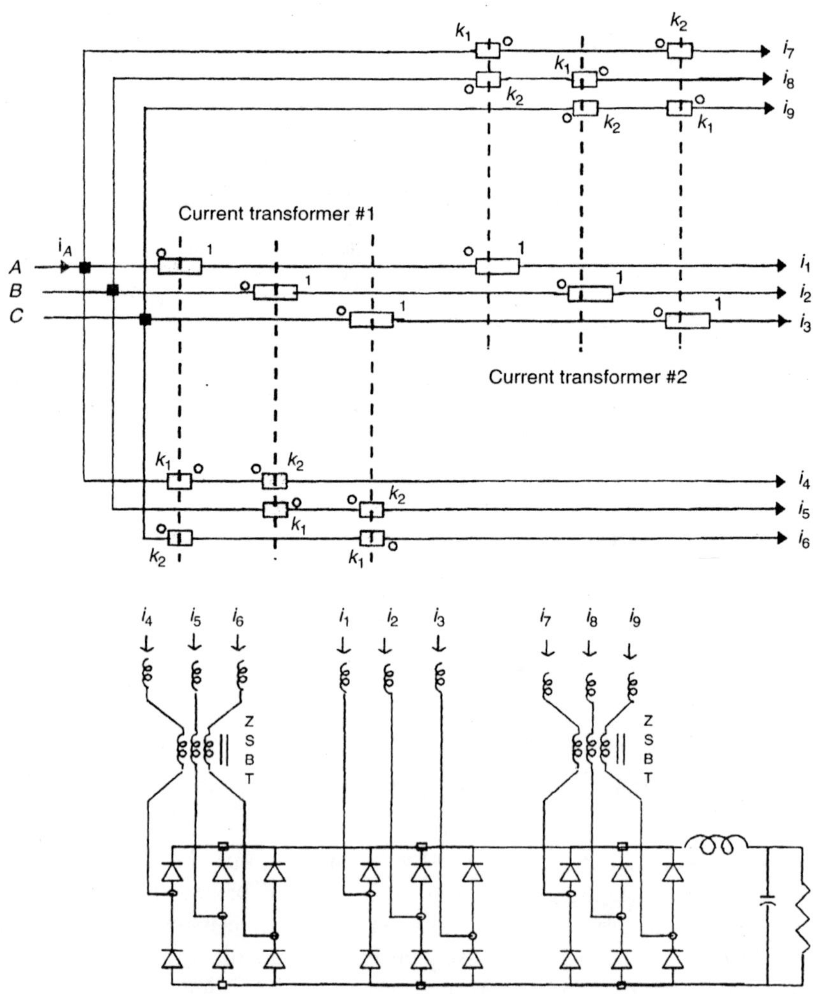

Figure 6-17 Conceptual 18-pulse method using ZSBTs and current transformers.

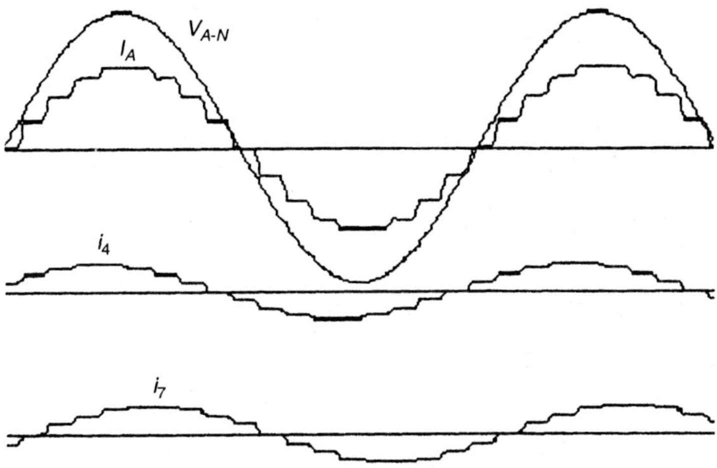

Figure 6-18 Computer results from 18-pulse circuit in Figure 6-17.

schemes discussed by Depenbrock and Nieman [6]. Nevertheless, it shows what can be done with current transformers.

In general, the blocking transformer (line interphase transformer) is required to absorb different voltages, including fundamental, zero sequence, and harmonics. The circuit requires a certain amount of ac line reactance.

Practical arrangements, without separate phase-shifting transformers, have been described by Depenbrock and Nieman for 12-pulse and 18-pulse [6]. Their connections incorporate turns that use the total line current in addition to the individual converter currents. This results in very efficient use of the transformer iron and copper.

As described in their papers, a 12-pulse circuit requires a three-phase ac line inductor with an equivalent VA rating of about 7% of the load. Also needed are the three current transformers (line interphase transformers). These have a combined equivalent rating of 13.4% of the load. Thus, the total magnetic parts are equivalent to 20.4% of the load.

In these circuits the components are very well utilized. Each of the 12 diodes conduct current for nearly 180°. There is no dc filter inductor, just a large capacitor.

In one practical 12-pulse design, the output voltage was about 11% less than that obtained by straight-through rectification. This would probably be too low for most variable-frequency drive systems operating from 480 V. Additionally, the lack of dc filter inductor makes the design rather sensitive to ac line voltage unbalance. However, design variations can address these limitations.

Chapter 7

Calculation of Harmonics

7.1 INTRODUCTION

Conforming with harmonic requirements specified in IEEE Std 519–1992 requires calculation of both voltage and current harmonics. This chapter describes how to make these calculations in a straightforward manner with appropriate engineering accuracy. The results can rarely be exact due to numerous system and component variations that influence the result.

In the IEEE specification, total harmonic distortion expresses the total harmonic voltage or total harmonic current as a percentage of the total fundamental components. Thus,

$$\text{total distortion of voltage} = THD_v = \frac{V_H}{V_1} \times 100\% \tag{7.1}$$

$$\text{total distortion of current} = THD_I = \frac{I_H}{I_1} \times 100\% \tag{7.2}$$

where V_1 is the system fundamental component of nominal voltage and I_1 is the system fundamental component of current. (This includes current drawn by both linear and nonlinear loads.)

$$V_H = \sqrt{\sum\nolimits_{h=2}^{h=\infty} V_h^2} \tag{7.3}$$

$$I_H = \sqrt{\sum\nolimits_{h=2}^{h=\infty} I_h^2} \tag{7.4}$$

In most practical cases, good results can be obtained by limiting the upper level of summation in the preceding equations to $h = 25$. If there is substantial "notching," as in some thyristor circuits, it is desirable to check whether significant higher-frequency harmonics are present.

7.2 CALCULATION OF HARMONIC CURRENTS

Individual current harmonics are under the control of the equipment manufacturer and can be modified by changing the equipment design. Practical variations in topology range from a simple three-phase bridge rectifier with 6-pulse current harmonics to 18-pulse and higher converter systems. The 6-pulse converters cause current and voltage harmonics that have frequencies of $(6k \pm 1)$ and amplitudes of $1/(6k \pm 1)$. Similarly, the frequency and amplitude factors of 12-pulse are $(12k \pm 1)$ and of 18-pulse $(18k \pm 1)$.

The converter equipment will draw amplitudes of ac line harmonic currents that fall into these characteristic patterns, providing that the dc current is fairly smooth. To achieve smooth dc current with most practical loads, the dc circuit must incorporate an appropriate filter inductance. This filtering is sized not just to limit the effects of the anticipated characteristic ripple voltage, but also the effects of line voltage unbalance and preexisting harmonic voltages. The latter becomes more evident in multipulse converters. For 60-Hz supply systems, good results are obtained in the presence of type 2 power if the dc circuit inductance is defined by $\omega L_{dc}/R_{dc} \leq 0.084$ at full load, where

ωL_{dc} = the reactance of the dc inductance at 60 Hz
R_{dc} = the equivalent resistor corresponding to full load

For some loads, such as small pulse-width-modulated (PWM) inverters, it may not be necessary to limit the harmonics to those obtained with fairly smooth current. In this case, it is feasible to substitute about 2.5% ac line reactance for the dc filter. Conditions favorable to this approach include

- Total nonlinear load is small relative to the power system.
- The source is balanced type SP1 power (difficult to prove).

Without adequate dc or ac filtering, the harmonic currents are practically impossible to predict.

The data presented in this section will address use of both ac circuit and dc circuit filter inductance. The distinction between inductor-input and capacitor-input filters relates to the way the rectified output is fed into the dc load. Examples are given in Figure 7-1.

Where harmonic current concerns must be addressed, the use of an inductor-input filter offers significant advantages. The inductance provides control of low-frequency harmonics, such as the 5th harmonic, with a minimal dc voltage drop.

In some of the simplest low-power converters that have only a capacitor-input filter, some amount of ac line reactance is essential. If this reactance is significant, then the corresponding reduction in dc output voltage must be recog-

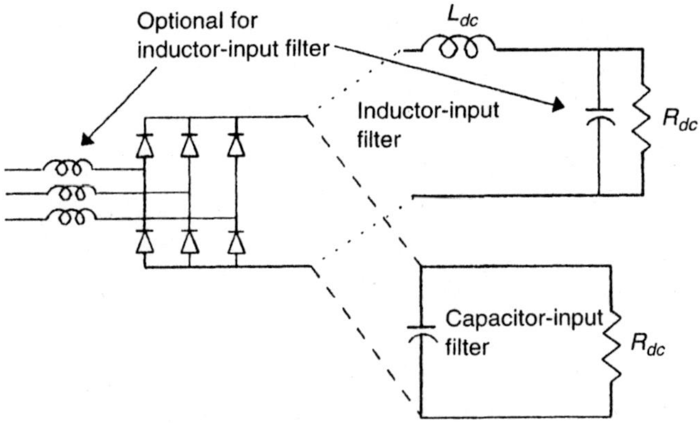

Figure 7-1 Inductor-input and capacitor-input filters.

nized. In practice, the resulting low-frequency current harmonics in these circuits, such as the 5th harmonic, are likely to be twice as large as those obtained when a dc inductor is used.

7.2.1 Accommodating the Effects of Source Unbalance

It is recommended to calculate converter harmonic currents on the assumption of a power source type 2. This includes 1% negative-sequence voltage; therefore, the line currents will not be balanced. This results in two effects that must be considered, namely,

1. Presentation of harmonic current data
2. Factors affecting 3rd harmonic current

These issues are now addressed.

7.2.2 Presentation of Harmonic Current Data

An appropriate mean value of harmonic currents is determined that reflects the heating effects. The way in which this is done is shown in Figure 7-2.

From this figure it is seen that an equivalent harmonic current is defined by

$$I_{eq} = \sqrt{\frac{I_A^2 + I_B^2 + I_C^2}{3}} \tag{7.5}$$

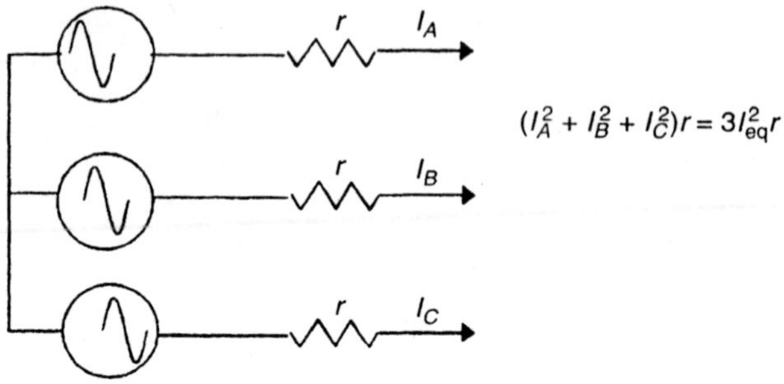

Figure 7-2 The method for calculating equivalent harmonic currents.

This is the definition that has been used in tabulating harmonic current data.

7.2.3 Effects of Circuit Reactance on 3rd Harmonic Current

This is one of those calculations that is extremely difficult to develop with a closed form solution but is straightforward with a computer simulation. Results of a computer simulation are shown in Figure 7-3.

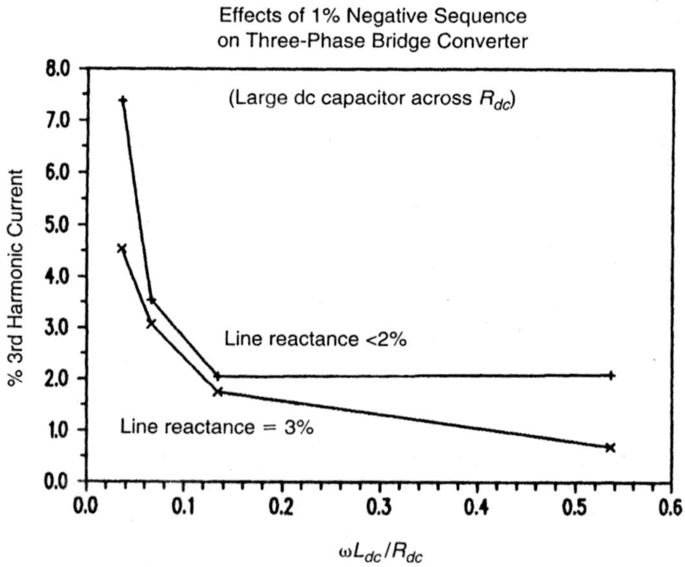

Figure 7-3 Effects of ac and dc circuit reactance on 3rd harmonic (1% negative-sequence source voltage).

Both ac and dc circuit inductance reduce the sensitivity of the 6-pulse (and other) circuits to ac source voltage unbalance. This further supports the recommendations to use a dc filter inductance when harmonic currents are to be controlled.

7.2.4 Effects of dc Filter Inductance

Because the filter inductance in a practical circuit is finite, there will be a certain amount of ripple current in the dc circuit. This ripple current modifies the percentage of various harmonic currents in the converter ac line current. Dobinson has evaluated this effect and documented the results, assuming negligible ac line reactance [7].

It is shown that in a 6-pulse bridge converter the 5th harmonic of current is increased when the ripple current increases, whereas other harmonics are reduced. For example, if the peak-to-peak ripple current I_r is defined relative to the direct current I_d, then the ac line 5th harmonic of current is given by

$$I_5\% = 20\left(1 + 0.945\frac{I_r}{I_d}\right) \tag{7.6}$$

Thus, if the peak-to-peak dc current ripple is 20%, the 5th harmonic of ac line current will be 23.8%.

In many practical circuits the dc output is provided by a rectifier with inductor-input filter, followed by a large dc filter capacitor. The resulting waveshapes are shown in Figure 7-4 for a 6-pulse rectifier.

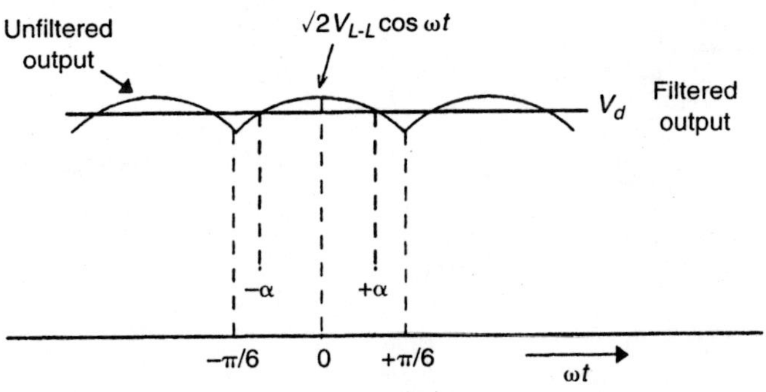

Figure 7-4 Waveshapes for a 6-pulse rectifier output.

As indicated in this figure, the volt-seconds to be absorbed by the filter inductor are given by

$$\text{inductor volt-secs} = \int_{-\alpha}^{+\alpha} \left(\sqrt{2}\,V_{L-L}\cos\omega t - V_d\right) dt$$

Now

$$\cos\alpha = \frac{3}{\pi}; \quad \therefore \ \text{volt-secs} = 0.025\frac{V_{L-L}}{\omega} \tag{7.7}$$

Since

$$V_d \approx \sqrt{2}V_{L-L}\frac{3}{\pi} \quad \text{and} \quad \frac{\omega L_{dc}}{R_{dc}} = 0.084$$

we get

$$\frac{I_r}{I_d} = 0.225$$

Incorporating this value of I_r/I_d into equation (7.6) gives a result of 24.3% for the 5th harmonic of current in the converter ac line, when $\omega L_{dc}/R_{dc} = 0.084$. This neglects ac line reactance, which, in practice, has only a small effect on the 5th harmonic of current, as observed in Figure 7-5.

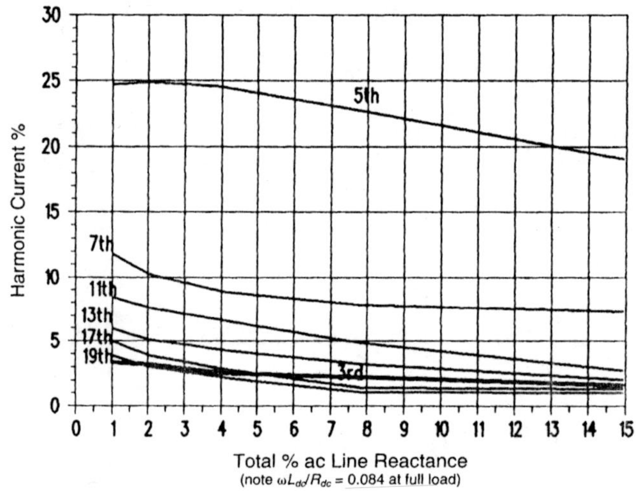

Figure 7-5 Harmonic currents of three-phase rectifier bridge with dc filter inductor (power type SP2).

7.3 CURRENT HARMONICS IN CONVERTERS WITH AN INDUCTOR-INPUT FILTER

For calculation purposes, the harmonic currents drawn by a three-phase rectifier bridge have been calculated from digital simulation programs. The results, which are displayed in Figure 7-5 and approximated by the equations in Table 7-1, are representative of practical designs. If initial system calculations show marginal adherence to the specifications, then more detailed digital simulation and calculation are recommended. If a dc filter capacitor is connected across the dc load as in variable-frequency-device (VFD) systems, a resonance frequency occurs. For most practical values, this resonance occurs at a frequency below the 2nd harmonic, and the capacitor has little effect on the curves of harmonic current.

TABLE 7-1.

THREE-PHASE BRIDGE RECTIFIER CURRENT HARMONICS WITH TYPE 2 POWER, $\omega L_{DC}/R_{DC} = 0.084$

$$\% \text{ harmonic current} \approx A_I e^{(B_I X_{pct})}$$

where

$$X_{pct} = 100\frac{I_{FL}}{I_{SC}} = \% \text{ ac reactance at rectifier terminals}$$

Harmonic #	A_1	B_I
3	3.35	−0.0515
5	25.9	−0.0194
7	10.9	−0.0308
11	9.1	−0.0807
13	6.0	−0.0747
17	4.4	−0.0946
19	3.42	−0.097
23	2.7	−0.11
25	2.3	−0.12

7.3.1 General Multipulse Considerations

For practical reasons, the harmonics that are expected to be canceled by multipulse connections are not completely eliminated. Different manufacturers will achieve different results depending upon their designs.

Where specific data is unavailable, initial multipulse calculations can be made on the assumption that residual harmonic currents have values that are 20% of those defined for 6-pulse converters. (See Figure 9-8.)

For example, in an undefined 12-pulse converter, make an initial assumption that the 5th, 7th, 17th, and 19th harmonic currents will be 20% those of a similarly rated 6-pulse converter.

7.4 CURRENT HARMONICS IN CONVERTERS WITH CAPACITOR-INPUT FILTER

Digital simulations are again used to develop harmonic current data as a function of ac line reactance. Because of the sensitivity of this type of circuit to unbalance and preexisting harmonic voltages, the data is provided for power type SP1 only.

Some observations on the effects of unbalance and preexisting harmonic voltages are made as follows:

- With a total of 2.5% ac line reactance, 1% negative-sequence voltage produces about 15% 3rd harmonic current. This is practically independent of the phase angle of the negative-sequence voltage. (In contrast, note that a dc filter inductor allows only 3% 3rd harmonic current.)
- Depending upon its phase angle, a 2.5% preexisting 5th harmonic in the source voltage may reduce or increase the 5th harmonic of current by about 15%.
- The presence of 5th harmonic voltage as well as negative-sequence voltage increases the amount of 3rd harmonic current.

Figure 7-6 and Table 7-2 result from fitting curves to the digital calculations. This data will help with system design when the converter has only a dc filter capacitor. To accommodate practical values of ripple current, the filter ca-

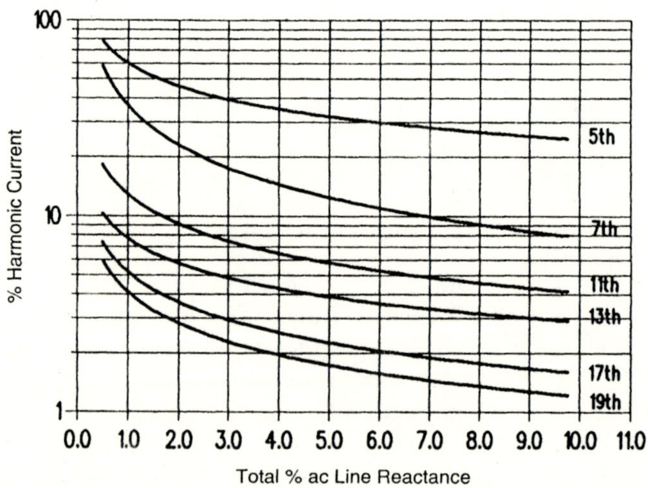

Figure 7-6 Harmonic currents of a three-phase rectifier bridge with dc filter capacitor and power source type SP1.

pacitance will be large when the load is a variable-frequency inverter. For example, 100 μF/HP would be typical in 480-V systems. Capacitors ranging from 50 μF/HP to 500 μF/HP do not detract from the practical usefulness of the curves. For much smaller capacitance, such as less than 10 μF/HP, additional digital simulation is desirable.

TABLE 7-2.

THREE-PHASE BRIDGE RECTIFIER CURRENT HARMONICS WITH TYPE 1 POWER (INPUT-CAPACITOR FILTER)

$$\% \text{ harmonic current} \approx A_{IC}(X_{pct})^{B_{IC}}$$

where

$$X_{pct} = 100\frac{I_{FL}}{I_{SC}} = \% \text{ ac reactance at rectifier terminals}$$

Harmonic #	A_{IC}	B_{IC}
Approx. 3rd with 1% negative sequence	32	–0.87
5	59.6	–0.386
7	36.38	–0.67
11	12.87	–0.5
13	7.67	–0.42
17	5.14	–0.512
19	4.08	–0.531
23	3.0	–0.634
25	1.71	–0.40

7.5 CALCULATION OF SYSTEM VOLTAGE DISTORTION

Calculation of voltage distortion requires knowledge of the harmonic currents. To calculate these, it is convenient to treat power electronics equipment as a harmonic current generator. This assumes that the effects of preexisting voltage distortion are not large and that the harmonic currents are defined for the most part by the equipment design topology. This is further discussed in Section 8.7.3. Harmonic voltages produced by the harmonic currents depend upon the source impedance.

A typical system one-line representation is shown in Figure 7-7. Two possible points of common connection are shown because different users may require calculations at different points. The system is further reduced to a simple per-phase representation. This facilitates calculations in which the individual harmonic currents are summed and then used to develop individual harmonic voltages. From these, total voltage distortion can be determined.

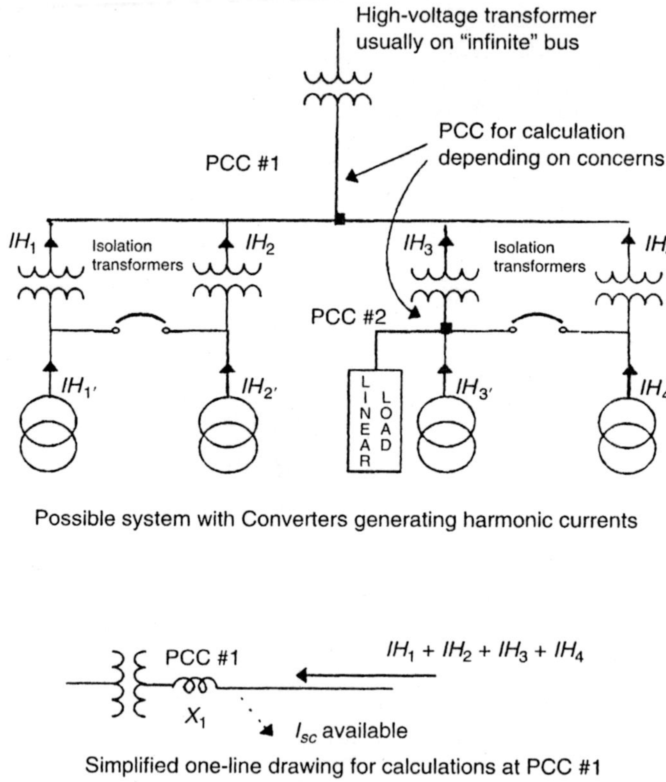

Figure 7-7 System drawing for harmonic calculations.

For power systems incorporating parallel ac line capacitors, the effective source impedance may vary significantly with frequency. A resonance (high impedance) occurs at harmonic number h_r, which depends upon the available short-circuit capacity (kVA_{sc}) and capacitor rating ($kVAC$), as shown in equation (7.8).

$$h_r = \sqrt{\frac{kVA_{sc}}{kVAC}} \tag{7.8}$$

If h_r is near to the converter current harmonics ($6k \pm 1$), current magnification may occur. This requires detailed design analysis.

In many applications the source can be represented by a simple reactive impedance. In this event the calculations are greatly simplified. A simple reactive impedance X_1 is assumed in the ensuing discussions. The impedance is determined based on the available short-circuit current capability I_{sc} and the line-to-neutral voltage V_N. In this case we obtain X_1 equals V_N/I_{sc}.

For harmonic calculations, this fundamental frequency impedance is assumed to be purely reactive. Thus, if a 480 V source has a short-circuit capability of 20 kA, its effective reactance at 60 Hz is 277/20,000, that is, 0.01385 Ω per phase. A 5th harmonic of current will see this as an impedance that is five times greater, that is, 0.06925 Ω per phase.

The harmonic voltage V_h developed across the fundamental frequency source impedance X_1 by a harmonic current I_h is given by

$$V_h = I_h h X_1 \tag{7.9}$$

where h is the harmonic number.

Applying the definition for total harmonic voltage distortion given in (7.1) results in

$$THD_V = \frac{X_1}{V_1}\sqrt{\sum_{h=2}^{h=\infty}(I_h h)^2} \times 100\% \tag{7.10}$$

where

$$\frac{X_1}{V_1} = \frac{1}{I_{sc}}$$

the reciprocal of the short-circuit current.

It is helpful to examine (7.10) to determine whether any generality can be achieved. We would expect that the converter harmonic currents could be expressed relative to the fundamental current (load), but other factors are involved, such as the type and amount of dc filtering and the total percentage of ac line reactance.

Since we have previously characterized the two types of dc filter as being inductor-input or simply capacitor-input, we need data for these two types. The total ac line reactance is an important part of the equation also.

To get a general approach we need a function that includes I_h/I_1, X_{pct} [(X_{pct} = 100 I_1 X_1/V1)], and filter type. By defining a "% harmonic constant" as

$$H_c = 100\sqrt{\sum_{h=2}^{h=\infty}\left[h\left(\frac{I_h}{I_1}\right)\right]^2} \tag{7.11}$$

we obtain

$$\%\ \text{voltage distortion} = H_c \frac{I_{FL}}{I_{sc}} \tag{7.12}$$

where

I_{FL} = the converter full-load fundamental current
I_{sc} = the system short-circuit current at the point where the distortion is being calculated

The advantage of using a harmonic constant is that it is not necessary to repeat detailed calculations once the converter harmonics are defined. Similar converter topologies generate similar proportions of harmonic current. Therefore, the percentage harmonic constant can be described as a "distortion signature."

It is gratifying to find a very simple approach to a very complex calculation.

The harmonic constant H_c has a simple relationship to the k factor used in determining transformer derating when the total load on a transformer is nonlinear. Thus, the harmonic constant can provide system and transformer data. Use of H_c in determining transformer derating in the presence of harmonics is discussed further in Section 7.7.

Being a practical number, the harmonic constant is subject to a range of values caused by variations in specific equipment design, such as proportions of dc filter and ac filter. Nevertheless, it provides a simple and effective calculation technique.

Using the previously given harmonic current data, typical harmonic constants have been prepared. These are shown for three-phase bridge rectifier converters in Figure 7-8.

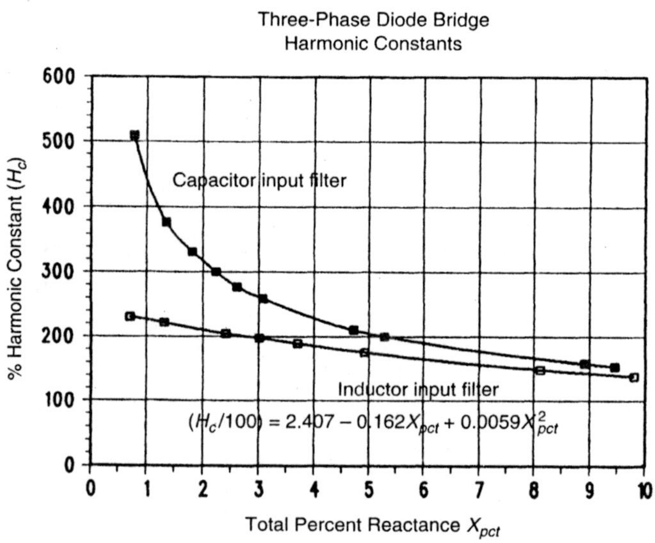

Figure 7-8 Harmonic constants for voltage distortion calculations with three-phase diode bridge converters.

The diode-type data in Figure 7-8 can be used to rapidly estimate whether a harmonic voltage distortion problem exists. If it does, then specific currents from the data in Tables 7-1 and 7-2 can be used to refine the calculations.

If further design changes are required, Table 7-3 provides a guide for selection of an alternative circuit topology. Once again, this data assumes a large value of dc filter capacitor.

To facilitate understanding of Table 7-3, consider the following:

- H_c tends to the lower values for diode circuits.
- "Conventional" refers to multipulse arrangements formed from parallel or series connection of 120° conduction, 6-pulse bridge connections.
- "Midpoint," here, indicates that half of the devices are in each commutating group; hence, each device conducts for a period of 720°/pulse number.

TABLE 7-3.

TYPICAL PERCENTAGE HARMONIC CONSTANT FOR THREE-PHASE CONVERTERS

H_c	Circuit Description and dc Filter
160 to 450	6-Diode C. filter
140 to 250	6-Diode/SCR L. filter
70 to 110	12-Diode/SCR L. filter, conventional
40 to 70	18-Diode/SCR L. filter, conventional
40 to 70	12-Diode/SCR L. filter, midpoint
22 to 50	18-Diode/SCR L. filter, midpoint

7.5.1 Application Examples

These examples focus upon VFD applications that represent a significant, and growing, source of installed converter load on the utilities. Harmonic calculation procedures are the same for other converter installations.

Unless noted otherwise in these examples, the following assumptions are made:

- Source power type SP2 exists with 2.5% preexisting 5th harmonic and 1% negative-sequence voltage.
- Total throughput efficiency, including losses in motor, VFD, cables, and so on, is 88%.

- Displacement power factor is 0.97.
- The utility system impedance is negligible compared to the transformer supplying the converter.

Example 1

Equipment—125-HP VFD with three-phase diode bridge having dc filter inductor and an additional ac line reactance of 2%. Power is supplied from a 500-kVA, 12-kV to 480-V transformer, with 5.75% impedance.

Calculation of Voltage Distortion at Transformer 480-V Bus—
VFD fundamental current I_1:

$$I_1 \approx 125 \times 746 \times \frac{1}{0.88} \times \frac{1}{0.97} \times \frac{1}{\sqrt{3}} \times \frac{1}{480} = 131.4 \text{ A}$$

transformer rated secondary current (without forced air cooling):

$$\frac{500 \times 1{,}000}{\sqrt{3} \times 480} = 601A$$

short-circuit current at point of calculation I_{sc}:

$$\frac{601}{0.0575} = 10{,}452 \text{ A}$$

transformer reactance referred to the *VFD*:

$$\frac{5.75\% \times VFD \text{ fl amps}}{\text{transformer fl amps}} = 5.75\% \times \frac{131.4}{601} = 1.26\%$$

total line reactance at VFD converter terminals:

$$2\% + 1.26\% = 3.26\%$$

From Figure 7-8 the percentage harmonic constant is 194; thus, calculated total harmonic distortion of voltage = THD_V:

$$\frac{194 \times 131.4}{10{,}452} = 2.44\%$$

The calculation of voltage distortion is straightforward when only a single nonlinear load is on the system.

Example 2

Using the same power source, another VFD is to be added rated at 250 HP. The total transformer loading is now 375 HP. As a first step, this 250-HP VFD will have a dc filter inductor, 2% additional ac line reactance, and generally the same harmonic signature as the 125-HP unit. The total fundamental load current will now be 131.4 × 375/125, that is, 394 A.

Because transformer reactance is now a bigger percentage with respect to the nonlinear load, the harmonic constant will be reduced. transformer reactance referred to total VFD load:

$$\frac{5.75 \times 394}{601} = 3.77\%$$

total line reactance at equivalent VFD terminals:

$$3.77\% + 2\% = 5.77\%$$

From Figure 7-8 the percentage harmonic constant is 167; thus, calculated total harmonic distortion of voltage is:

$$THD_V = \frac{167 \times 394}{10,452} = 6.29\%$$

This result does not meet the IEEE specification, so we must look for ways to reduce the voltage distortion. Various techniques are discussed in Chapter 8. One technique that we can evaluate with the data presented so far is to determine whether the use of a 12-pulse converter on the 250-HP drive would sufficiently reduce the total harmonic distortion.

This calculation can be approximated by calculating currents for the individual converters and then summing the results. The percentage reactance is different in each case because the referred transformer reactance depends upon the load.

For example, the referred transformer reactance is 1.26% for the 125-HP load and 2.52% for the 250-HP load. Two percent additional reactance within the equipment must be added to these to get the total percentage reactance of 3.26% and 4.52% for the 125 HP and 250-HP units, respectively. Using this data in conjunction with Table 7-1, the individual harmonic currents can be determined.

Calculated results are shown in Table 7-4. These results include individual converter currents and modified results for 12-pulse equipment, which is assumed to have residual uncharacteristic harmonics of 20%.

With this combination of 6-pulse and 12-pulse units, the harmonic voltage distortion is reduced to 4.86%, and the IEEE requirements are met.

It will be shown later in Chapter 8, Section 8.5, that similar improvements can be obtained, at full load, if the 250-HP drive is fed from a transformer with a 30° phase shift. Many numerical calculations are required to obtain these results. Clearly, the procedure can be greatly simplified with dedicated computer software. Some manufacturers routinely use this type of computer program.

TABLE 7-4.

HARMONIC CURRENTS AND VOLTAGES FOR 500-KVA, 480-V TRANSFORMER LOADED WITH 125-HP 6-PULSE AND 250-HP 12-PULSE VFDS

	125-HP 6-Pulse Amps	250-HP 6-Pulse Amps	250-HP 12-Pulse Amps	Total Amps	Harmonic Voltage
I_1	131.4	262.8	262.8	394.2	$V_N = 277.1$
I_3	3.72	6.96	6.96	10.68	0.85
I_5	31.9	62.4	12.47	44.37	5.88
I_7	12.9	24.9	4.98	17.88	3.32
I_{11}	9.2	16.6	16.6	25.8	7.52
I_{13}	6.2	11.3	11.3	17.5	6.03
I_{17}	4.24	7.54	1.51	5.75	2.59
I_{19}	3.27	5.81	1.16	4.43	2.23
I_{23}	2.47	4.31	4.31	6.78	4.14
I_{25}	2.05	3.52	3.52	5.57	3.69
total harmonic voltage = 13.48 V					
total harmonic voltage distortion = 4.86%					

7.5.2 Multiple Loads

For multiple nonlinear loads, a single equivalent load can sometimes be defined if the equipment has equal or nearly equal percentage reactance. This was done in the initial assessment for Example 2 earlier. If this is not possible, then the individual harmonic currents can be summed using the data in Tables 7-1 and 7-2. Voltage harmonics are then individually calculated to arrive at the voltage distortion. This technique was applied in the 6-pulse/12-pulse calculation in Example 2.

The technique of calculating individual converter harmonic currents and then summing to get the total harmonic current gives results that are somewhat higher than full digital simulations. There are two main reasons for this:

1. The source reactance referred to individual loads is lower than for the total load. This causes calculations of individual converter currents to be higher than actual.
2. Practical differences in converter designs cause phase shifts in the harmonic currents such that some cancellation of harmonics occurs. This reduction in harmonics is more likely in the higher harmonic frequencies.

The conservative nature of results calculated by these means may be helpful in compensating for numerous unspecified practical variations. Some practical concerns about measurements are given in Chapter 8, Section 8.7.

If the multiple equipments are of different design and have mixed percentage reactances, the same initial approach can be used. However, these simple calculations may be unnecessarily conservative. If results are marginal, a digital solution is recommended for more accurate results when multiple mixed-design loads are deployed.

7.6 CALCULATION OF SYSTEM CURRENT DISTORTION

As noted in Section 7.1, current distortion expresses harmonic currents relative to total demand current I_L, which includes the fundamental current of nonlinear and linear equipment. A current distortion percentage can easily be defined for any particular nonlinear load, but this may not be useful for the complete system. The calculation of total current distortion can be made using the data in Tables 7-1 and 7-2 in conjunction with known system loading.

7.7 TRANSFORMER *K* FACTOR AND H_C RELATIONSHIP

Specification ANSI/IEEE Std C57.110-1986 describes a technique for calculating derating of standard transformers in the presence of harmonic currents. In this method the harmonic currents must be expressed relative to the total fundamental current. This procedure is similar to the technique used in IEEE Std 519–1992 for defining current distortion. A special case arises when the only load on the transformer is nonlinear.

In applying the formulas in the ANSI/IEEE specification, it is useful to first develop a factor that relates to the harmonic and fundamental currents. The transformer industry refers to this as the *k* factor. It is defined here in equation (7.13),

$$k = \frac{\sum_{h=1}^{h=h_{\max}} h^2 (I_h/I_1)^2}{\sum_{h=1}^{h=h_{\max}} (I_h/I_1)^2} \tag{7.13}$$

where h is the harmonic number and (I_h/I_1) is the ratio of harmonic current to fundamental current.

A special case arises when the only load on the transformer is nonlinear, because in this case a simple relationship exists between the "*k* factor" and the "harmonic constant"; namely,

$$k = \left[1 + \left(\frac{H_c}{100}\right)^2\right]\left(\frac{I_1}{I_{\text{rms}}}\right)^2 \tag{7.14}$$

where I_1/I_{rms} is the ratio of fundamental to rms current.

We can use these formulas to calculate the transformer current capability in the following way.

Let the winding eddy-current loss under rated conditions, expressed per unit of rated load I^2R loss, be $P_{EC\text{-}R}$ (pu).

Then from ANSI/IEEE Std C57.110-1986, the transformer maximum per-unit root-mean-square (rms) current rating $I_{\max}$ (pu) is established as

$$\text{transformer } I_{\max}(pu) = \sqrt{\frac{1 + P_{EC-R}(pu)}{1 + k\, P_{EC-R}(pu)}} \tag{7.15}$$

Typically, $P_{EC\text{-}R} = 0.15$.

When the total load on the transformer is nonlinear, we can use the harmonic constant directly to determine k and hence the transformer derating.

When the load is mixed, the data in Tables 7-1 and 7-2 can be used, in conjunction with system load currents, to determine the k factor.

7.8 SINGLE-PHASE POWER CONVERSION

Individual 120-V equipment is usually limited to a maximum line current of 15A. Thus, the line current of any particular equipment is not large enough to represent "power" electronics, at least not for the purposes of discussion in this book. Nevertheless, there may be large quantities of small equipment distributed across the power system. In total, these may represent an appreciable load. Harmonic effects created by this combined loading can be calculated from the anticipated harmonic currents of individual equipment. To facilitate these calculations, analysis of a representative equipment is given.

In some cases, the 120-V source is transformed and converted to a lower value of dc. In other cases, the power source may simply be rectified. In either case, the dc power is then probably processed by a switching regulator or other control means for use by the electronic circuits. A single bridge rectifier circuit, shown in Figure 7-9, can be used to develop basic harmonic data.

In many cases the dc inductor L_{dc} will be omitted for reasons of cost. The capacitor at the dc output affects the actual dc output voltage and the rating of any follow-on dc regulator. For practical values of filter capacitor, the ripple voltage is less than 10% peak to peak. A capacitor five times greater than necessary to achieve this has only a small effect on the harmonic currents. A full-load peak-to-peak voltage ripple of about 5% of the dc output is present in the circuit used to develop harmonic current data.

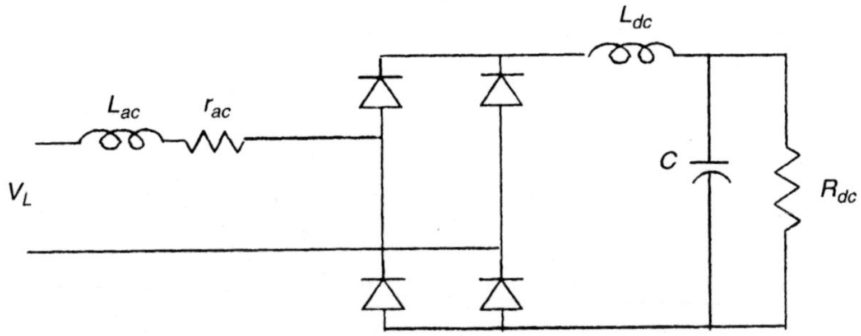

Figure 7-9 Basic single-phase bridge rectifier.

7.8.1 Effects of Circuit Resistance

In contrast with larger transformers, where leakage reactance is a significant factor modifying harmonic currents, a small transformer has significant winding resistance, for example, 4% for a 130-W supply. This resistance reduces harmonic currents to some extent, as shown in Table 7-5. In this table the ac line reactance and dc circuit inductance are negligible. Percentage resistance is the ratio of I_1/I_{sc}, where I_1 is the ac line fundamental current and $I_{sc} = V_{line}$/resistance.

The single-phase circuit produces a large amount of 3rd harmonic current. This contrasts with three-phase converter circuits in which 3rd harmonic currents are developed only by reason of practical limitations, such as unbalance.

TABLE 7-5.

SHOWING THE EFFECT OF SOURCE RESISTANCE ON A SINGLE-PHASE RECTIFIER BRIDGE WITH NOMINAL 5% PEAK-TO-PEAK OUTPUT RIPPLE		
% resistance	**$\%I_3$**	**$\%I_5$**
0.92	93.6	81.8
1.77	91.1	75.1
3.42	86.1	62.5
6.4	79	46.6

7.8.2 Effects of Source Reactance

Harmonic currents can be reduced to some extent by adding ac line reactance. In practice the distribution transformer provides some reactance, which, although negligible for any one rectifier circuit, may affect a combination of circuits. To determine the effectiveness of ac reactance, design data is given in Figure 7-10. In developing this data, a source resistance of 0.9% was included. Peak-to-peak output ripple voltage was 5.3% for less than 1% reactance and was reduced to 3.2% for 11.4% reactance.

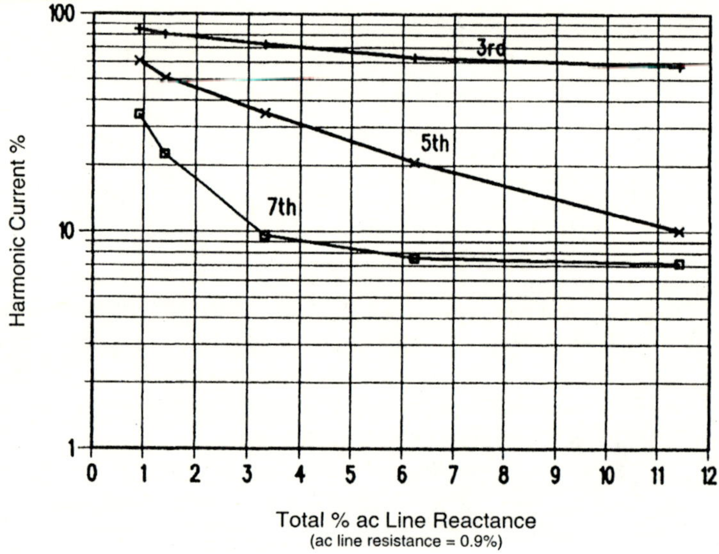

Figure 7-10 Showing the effect of ac line reactance on a single-phase rectifier bridge with dc capacitor-input filter.

7.8.3 Effects of 3rd Harmonic Currents

Depending upon the power system arrangement, the 3rd harmonic of current, associated with single-phase nonlinear loads, may significantly raise the rms current in the neutral conductor and supply transformer. A possible power system arrangement is shown in Figure 7-11.

The neutral fundamental current can be reduced by balancing the single-phase loading. However, for nonlinear loads, the 3rd and other triplen harmonic currents are almost in phase (depending upon ac line reactance). These harmonics thus add almost arithmetically in the neutral conductor, and must be accommodated in the system design. For current pulses of less than 60° the rms current in the neutral is $\sqrt{3}$ times the rms current drawn by balanced nonlinear loads. This result can also be used for practical values of the source ac line reactance, which may extend current conduction.

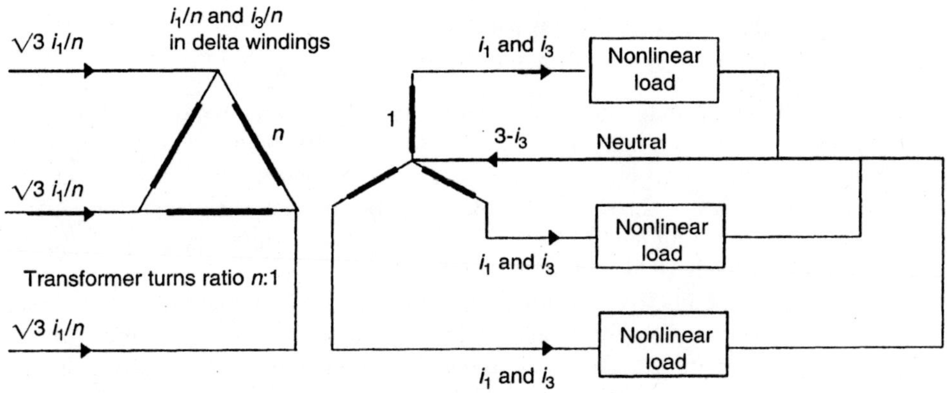

i_1 and i_3 are the amplitude of fundamental and 3rd harmonic currents, respectively.

Figure 7-11 Possible power distribution for single-phase loads.

In a practical arrangement, neither the fundamental nor the 3rd harmonic of current will be equal in each nonlinear load. Therefore, some degree of unbalance will exist in the system ac source. Some 3rd harmonic current will circulate in the delta winding; the remainder will flow as an unbalanced 3rd harmonic of current in the source. Currents in the case of severe unbalance, with a single nonlinear load, are shown in Figure 7-12. Other power systems can be analyzed in a similar manner.

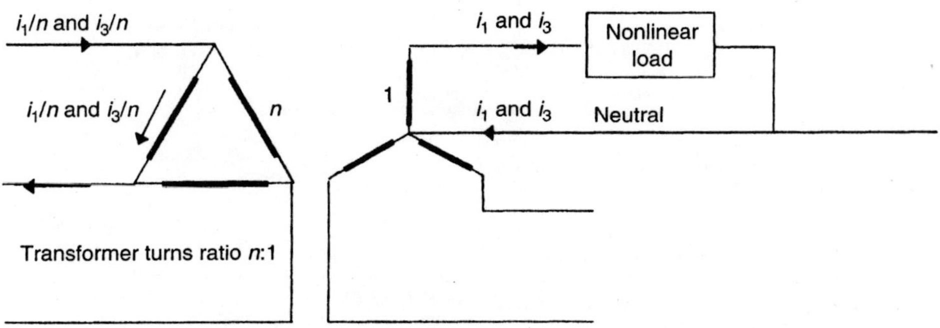

i_1 and i_3 are the amplitude of fundamental and 3rd harmonic currents, respectively.

Figure 7-12 Current division with load on only one phase.

7.8.4 Reducing Harmonics of Single-Phase Rectifiers

System benefits can be obtained by reducing the 3rd harmonic of current associated with nonlinear single-phase circuits. Possible circuits to achieve this end include tuned filters on the ac side, dc filter inductor before the capacitor, and active filter circuits. As shown in Figure 7-13, the dc filter inductor is very effective in reducing the 3rd harmonic of current. With this type of filter, the 3rd harmonic is easily reduced to less than 40%.

Active circuits, which involve high-frequency pulse-width modulation methods, can provide significant reduction of all the major harmonic currents. These methods are well described in the literature and will not be further investigated here [26, 27].

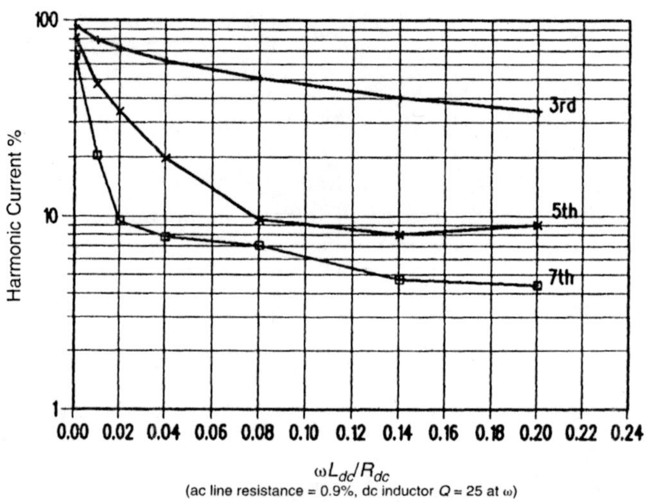

Figure 7-13 Effect of L_{dc} on the single-phase rectifier bridge shown in Figure 7-8.

7.9 EFFECTS OF USING PHASE-CONTROLLED THYRISTORS

Most of the data derived so far has focused on the use of different topologies, with different power source characteristics, but with diode rectifiers. Thyristors find little application with single-phase converters, but have significant use with three-phase circuits. The impact of thyristors on harmonic currents and voltage distortion associated with three-phase topologies will now be discussed.

Phase control delays the moment at which commutation is allowed to occur, and this causes a higher voltage to be applied to the circuit ac inductance. The result is a faster rate of change of device and line current as the current transfers from one device to the next. This faster switching action generates more high-frequency current harmonics and notching effects. A discussion of the notching effects produced by phase-controlled thyristors is given in Chapter 1, Section 1.2.

Phase control also causes a higher ac ripple voltage to appear across the circuit dc inductance; however, a larger (more expensive) dc inductor can be selected to control the amount of peak-to-peak ripple current.

In practice the substitution of phase-controlled devices results in greater amounts of harmonic current, especially those of higher frequency. Table 7-3 indicates that the harmonic constant and hence line voltage distortion may increase by a factor of 1.5 to 2.2 times when phase controlled devices (SCRs) are deployed. These effects are undesirable; however, Figure 10-4 shows that excellent results can be obtained in multipulse circuits, providing ac and dc inductance values are sized appropriately.

7.9.1 Six-Pulse Circuits

Figure 7-14 compares the amplitudes of harmonic currents in one particular 6-pulse converter design with 0° and 60° phase control. In this example the dc filter inductor was changed to keep the same amount of ripple current in the dc circuit. The increase of higher-frequency harmonic currents is in accord with notch effects.

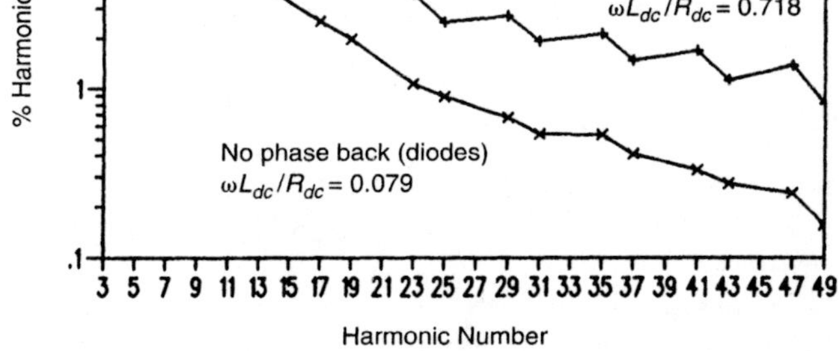

Figure 7-14 Six-pulse converter harmonics with 29% peak-to-peak dc ripple current. Source reactance = 4.4%

Instead of keeping fixed dc current ripple we can simply allow higher ripple current when the thyristors are phased back. In this case the effects can be judged from Figure 7-15.

It is noted that with a large amount of ripple current, a significantly larger 5th harmonic of current occurs, as predicted in equation (7.6); however, the high-frequency components are reduced. Reduction of the higher-frequency components is in line with reduced notching when the dc circuit ripple is larger. This is because the current at the instant of commutation is smaller.

The results shown in Figures 7-14 and 7-15 are representative of the performance expected in 6-pulse systems.

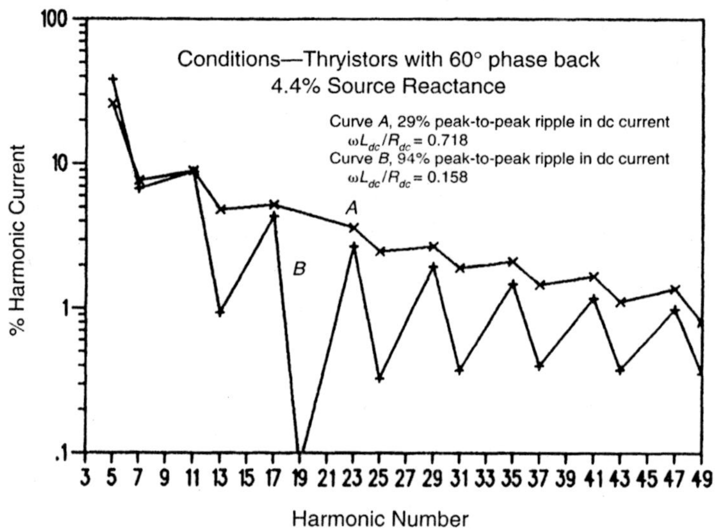

Figure 7-15 Six-pulse converter harmonics with 60° phase-back and different dc circuit ripple current.

7.9.2 Multipulse Circuits

When thyristors with phase-back control are used in multipulse circuits, the harmonic currents are increased. However, as shown in Figure 10-4, excellent low harmonic performance can be achieved in these circuits, by appropriate selection of components. Notching is especially reduced in midpoint multipulse methods. As explained in Chapter 9, Section 9.2, this is because the reduced commutating voltage, in conjunction with a small amount of phase reactance, greatly reduces the rate of change of current at commutation.

Appropriate component parameters will depend upon the specific multipulse topology, but the physical size of the components required to overcome phase-back effects is not large. For example, only 3.5% ac line reactance was used in one circuit with 18 thyristors. For specific designs a complete digital simulation is recommended.

Chapter 8

Meeting Harmonic Standards

8.1 INTRODUCTION

In Chapter 7 we described how to calculate voltage and current harmonic distortion for various types of equipment. Now we need to know what remedies are available when the calculated results are unsatisfactory. Techniques for meeting harmonics specifications will be described. The amount and types of power electronics equipment that can be deployed on different power systems, without exceeding IEEE Std 519-1992, are reviewed.

8.2 LIMITING INSTALLED NONLINEAR LOAD RATING

A certain rating of nonlinear load is feasible without exceeding either the voltage or current distortion limits of IEEE Std 519-1992. The amount will depend upon the type of power electronics equipment and the type of power source.

Helpful guidelines can be established using the three-phase diode rectifiers that find wide application in the adjustable-speed drives industry. The methodology can then be extended to other types of equipment.

Even with the problem simplified to application of three-phase bridge rectifiers, we still need to recognize other practical variations. Power source variations include utility transformer and auxiliary generator. The two types of dc link filter are capacitor input and inductor input. These basic source and load combinations are shown in Figure 8-1.

8.2.1 Nonlinear Load Limits Constrained by Voltage and Current Distortion Limits

It has been shown that similar converter topologies generate similar proportions of harmonic current and exhibit a "distortion signature." This is characterized numerically with a percentage "harmonic constant." We will now apply

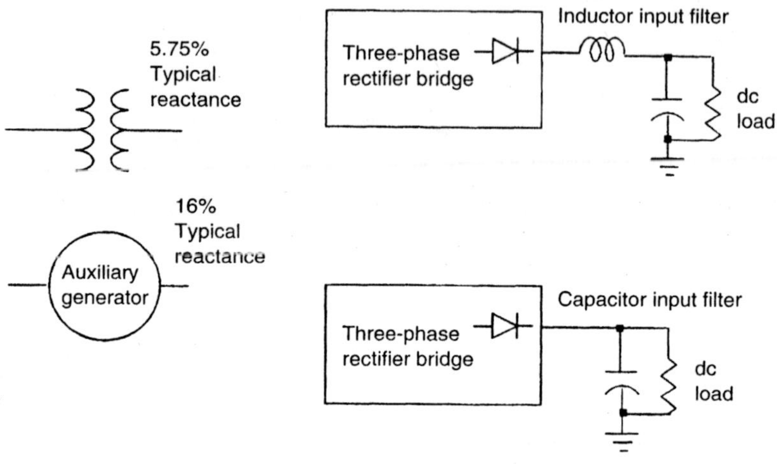

Figure 8-1 Basic power source and electronics equipment variations used with VFDs.

this concept to determine the maximum tolerable nonlinear loading on a power system.

From Figure 7-8, it is noted that rectifier bridges with an inductor-input filter produce a lower harmonic constant and hence less distortion than those with a simple capacitor filter.

Therefore, power systems can tolerate more nonlinear loading when the converters have an inductor-input filter. For this reason, this type of filter is expected to prevail as more power electronics load is installed.

The favorable harmonic features of the input-inductor filter will be used here to calculate the maximum amount of system nonlinear load without exceeding the IEEE requirements.

8.2.2 Converters (Without Additional ac Reactance) Fed from Utility Bus with Voltage Distortion as the Limit

From Figure 8-2, let I_{tr} be the transformer rated current (without forced cooling). Then let THD_V be the total voltage distortion—to be limited to 5.0%.

It is required to find the allowable ratio of rectifier load I_{fl} to the transformer rating I_{tr} without exceeding 5.0% voltage distortion. (Allowing for tolerances, the design goal will be 4.75%.)

Because the rectifier does not include additional ac line reactance, the short-circuit current at the rectifier terminals is the same as that of the transformer, namely, $I_{tr}/0.0575$; thus,

$$\frac{I_{sc}}{I_{tr}} = 17.39$$

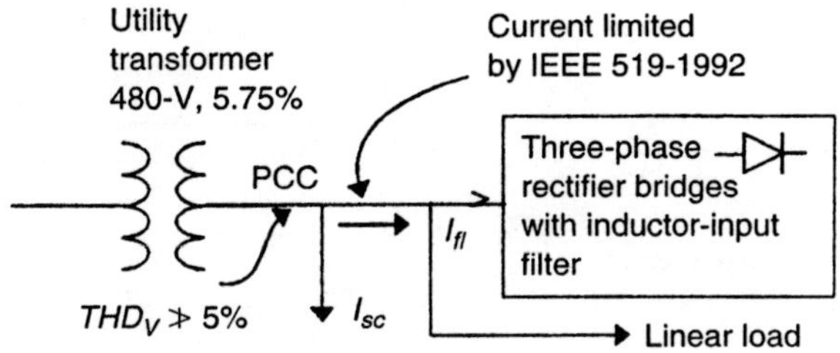

Figure 8-2 System schematic for utility-source calculation.

The effective percentage reactance of the transformer, referred to the rectifier, depends upon the full load ratings of rectifier and transformer. Thus, $X_{pct} = 5.75\ (I_{fl}/I_{tr})$.

With a 0.25% safety margin the total voltage distortion is then defined by

$$THD_V = 4.75\% = H_c \times \left(\frac{I_{fl}}{I_{sc}}\right) = H_c \times \left(\frac{I_{tr}}{I_{sc}}\right) \times \left(\frac{I_{fl}}{I_{tr}}\right) = H_c \times 0.0575 \times \left(\frac{I_{fl}}{I_{tr}}\right)$$

From Figure 7-8,

$$H_c = (100)(2.407 - 0.162X_{pct} + 0.0059X_{pct}^2) \tag{8.1}$$

where total percent reactance is given by

$$X_{pct} = 5.75 \times \left(\frac{I_{fl}}{I_{tr}}\right)$$

Substituting for X_{pct} in (8.1) and solving iteratively for THD_V, it is determined that

$$H_c = 206; \qquad \left(\frac{I_{fl}}{I_{tr}}\right) = 0.4; \qquad THD_V = 4.75\%$$

Thus, based on voltage distortion limits, the maximum tolerable three-phase rectifier load, using a typical utility bus with 5.75% transformer impedance, is 40% of the transformer rating. If ac line reactance of 3% is added to the rectifier, the allowable rating increases to 50%.

8.2.3 Converters Fed from Auxiliary Generator with Voltage Distortion as the Limit

Similar calculations are used for the generator arrangement, shown in Figure 8-3. The maximum tolerable three-phase rectifier load for auxiliary generators with 16% subtransient reactance is 14.4% of the generator rating. If line reactance of 3% is added to the rectifier, the allowable rating increases to 18%.

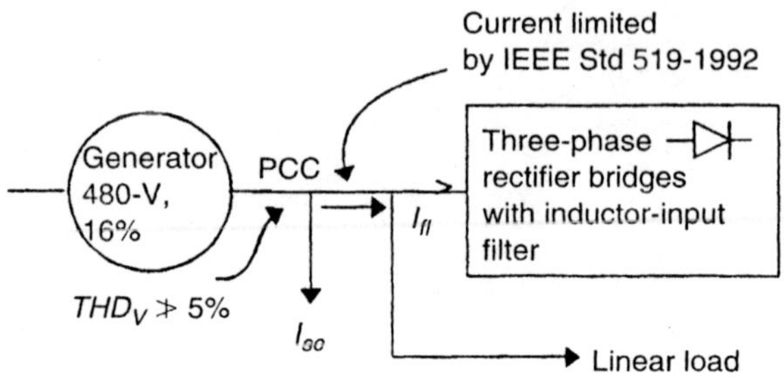

Figure 8-3 System schematic for generator-source calculation.

8.2.4 Converters Fed from Utility Bus with Current Distortion as the Limit

As described in Chapter 1, the allowable current harmonics are affected by the total system load current. Therefore, specific calculations are required for each case.

Despite this difficulty, two limiting cases are reviewed to get an appreciation of the effects of current distortion limits.

These two special cases are

Case (a): Only nonlinear load is present.

Case (b): Linear plus nonlinear load sums to the system rating.

There are numerous system variables that detract from the accuracy of a general solution. Therefore, in each case, it is necessary to calculate the line harmonic currents. Table 7-1 has already provided results that are representative of practical designs. These are used for analysis of the special cases being considered.

In the previous calculation, it was determined that the total nonlinear load had to be less than 40% of the transformer rating to meet voltage distortion limits. For current distortion calculations, we must first determine the short-circuit current ratio I_{sc}/I_L. The analysis is complicated here because different ratios apply depending upon whether case (a) or case (b) is being considered. For case (a) $20 < I_{sc}/I_L < 50$, and for case (b) $I_{sc}/I_L < 20$. Thus different amounts of 5th harmonic current are tolerable. From Table 1-2 these are given as 7.0% and 4.0%, respectively. With this preamble, we can proceed with the analysis of different scenarios.

Case (a) Assume the load consists entirely of three-phase rectifier bridges. Figure 7-5 shows that it is impractical to include additional ac line reactance to reduce the harmonic current to 7%. Other circuit topologies are available that would be satisfactory. These will be discussed later.

Case (b) Assume that the three-phase bridge rectifier load is limited to 40%. This is the maximum tolerable to meet voltage distortion. If the remainder of the system is fully loaded with 60% linear load, then the effective amount of 5th harmonic current becomes 23.5 × 0.4, that is, 9.4%. This is still not reduced enough to meet the current harmonic specifications. If the three-phase bridge rectifier load is further reduced to 29%, and the linear load is increased to 71%, then the effective 5th harmonic current will be 6.8%, and the current distortion limits will be met.

8.2.5 Converters Fed from Generator Bus with Current Distortion as the Limit

For auxiliary generators, nonlinear loading is restricted to 14.4% to allow the voltage distortion limits to be met. A similar amount (14%) is tolerable if the current distortion limits are to be met, provided there is also linear loading of about 70%. For example, a three-phase converter bridge has about 23.5% 5th harmonic current. If the nonlinear load is 14% and the linear load is 70% of the generator rating, then the effective 5th harmonic of current is given by

$$I_5 = \frac{14}{70+14} \times 23.5 = 3.9\%$$

More extensive general calculations appear to be of limited use. Treating each application on a case-by-case basis is recommended.

8.3 SUMMARY OF INSTALLED RATING LIMITS FOR THREE-PHASE BRIDGE RECTIFIERS

The foregoing discussions demonstrate that for utility power sources, meeting the IEEE Std 519-1992 current distortion limits is more difficult than is meeting the voltage distortion limits.

Additionally, simple uncorrected equipment, such as 6-pulse three-phase bridge rectifiers, are restricted in their deployment. In an example of utility application that was evaluated in a very favorable manner, the rectifier rating had to be restricted to about 29% of the total transformer load.

For an auxiliary generator, the limits of voltage distortion and current distortion require similar limitations in the rating of installed 6-pulse converters. In one application the nonlinear load was limited to 14% of the generator rating. Results have not been generalized, and it is recommended to make calculations for each individual case.

It is apparent from these deliberations that equipments with reduced harmonic currents, such as multipulse converters, can greatly increase the tolerable installed nonlinear load capacity. In fact with some 18-pulse equipments, discussed in Chapter 9, Section 9.7, the total utility or generator load can be nonlinear.

8.4 EQUIPMENT FILTERING METHODS

8.4.1 General

In addressing the means by which to reduce harmonic currents, it will be assumed that the converters incorporate a dc filter inductor such that the dc current is continuous. It has already been shown that this method of filtering gives fewer harmonics than does a simple dc filter capacitor.

8.4.2 Additional ac Line Reactance

The first approach to improving performance of individual three-phase 6-pulse converters is by way of adding ac line inductance. The effect of this can be judged quantitatively from the data, such as that given in Figures 7-5 and 7-6. Qualitatively, the inductor is responsible for slowing down the rate of rise when current transfers from one device to another (commutation). The amount of improvement attainable by this method depends upon the amount of dc voltage drop that can be tolerated under load.

For voltage drop calculations with a 6-pulse bridge converter, the ac line inductance can be represented by an equivalent resistor in the dc circuit of value $(\pi/6)X_{pu}$ [2]. However, note that there is no power loss in this equivalent series resistor. Over practical ranges, the reduction in output voltage is approximately half the percentage ac reactance. For example, an increase in reactance from 2% to 10% would reduce the output voltage under load from 0.99 V_d to 0.95 V_d. At the same time the harmonic constant would reduce from 210.6 to 137.7. This reduces voltage distortion by 34%. This amount of change may be sufficient in some cases. Certainly this is a simple method when the voltage drop, and hence increased current, can be tolerated.

Additional ac reactance can be provided by transformer leakage, but unless a double-wound transformer is required for other reasons, a separate ac reactor will be less costly and just as effective.

8.4.3 Multipulse Methods

In Chapter 3, the ability of phase-shifting transformers and multiple converters to produce multipulse effects was evaluated. The mechanism whereby harmonics are eliminated in pairs was explained. The minimum required phase shift for paralleled 6-pulse converters was defined in equation (3.1). When midpoint connections are used, the minimum phase shift is twice this value.

The characteristic line current harmonics are $(kq \pm 1)$, where k is any positive integer and q is the pulse number. Multipulse methods increase q; however, because of practical limitations, there will be small amounts of uncharacteristic harmonic frequencies present. Also the amplitudes of expected frequencies will be reduced due to reactance effects.

The harmonic constant for a conventional 12-pulse system, operating with type 1 power, is about one-half that of a 6-pulse system. Therefore, we can expect 12-pulse converters to nearly halve the system voltage distortion. This does not automatically double the amount of nonlinear equipment that can be deployed. Harmonic currents, especially the 11th and 13th, may now become the limiting feature. Other 12-pulse methods, such as that shown in Figure 9-12, have device conduction of 60° and experience greater filtering by ac line reactance. In these circuits the 11th and 13th harmonics are reduced, and a harmonic constant which is about 35% that of a 6-pulse system is feasible.

Specific multipulse circuits using phase-shifting transformers are evaluated in Chapter 9. In the 18-pulse configuration described there, current harmonics are reduced to the point where 100% of the load can be nonlinear.

8.4.4 Electronic Phase Shifting

The use of phase-shifting transformers and multiple converters as a means to reduce harmonic currents extends logically from a line of reasoning that attempts to reduce harmonic currents by adding the output of more than one converter. For example, the sum of two converters each with "ideal" current amplitude I_d, operating by some means at an angle Φ_n with respect to each other, is shown in Section 5.3 to be given by

$$i + i_n = \frac{4\sqrt{3}}{\pi} I_d \left[\begin{array}{l} \sin\left(\omega t - \frac{\Phi_n}{2}\right)\cos\frac{\Phi_n}{2} - \left(\frac{1}{5}\right)\sin 5\left(\omega t - \frac{\Phi_n}{2}\right)\cos\frac{5\Phi_n}{2} \\ \qquad - \left(\frac{1}{7}\right)\sin 7\left(\omega t - \frac{\Phi_n}{2}\right)\cos\frac{7\Phi_n}{2} + \cdots \end{array} \right] \tag{8.2}$$

If $\Phi_n = \pi/5$, that is, 36°, then the term involving the 5th harmonic will vanish. Likewise, a phase difference of $\pi/7$ will eliminate the 7th harmonic.

With this simple additive technique, we cannot eliminate the 5th and 7th current harmonics simultaneously; however, attenuation of other harmonics does result. For example, with $\Phi_n = 36°$, the 5th harmonic is eliminated, and the 7th harmonic, which is normally 14.2%, is reduced by a factor of $\cos 7\Phi_n/2$, to 8.4%.

When we try to set up an electronics circuit with an appropriate phase shift but without a transformer, difficulties arise that highlight the advantages of a phase shifting transformer. The phase shift could be obtained electronically by using two converters with phase control. As a compromise, a total shift of 30° can be used. This could be obtained with two converters. One with thyristors (SCRs) could be gated to operate at 15° lag. Another with gate-turn-off thyristors (GTOs) could be gated to operate with 15° lead. The two could then be combined through appropriate interphase transformers. A schematic of this possibility is shown in Figure 8-4.

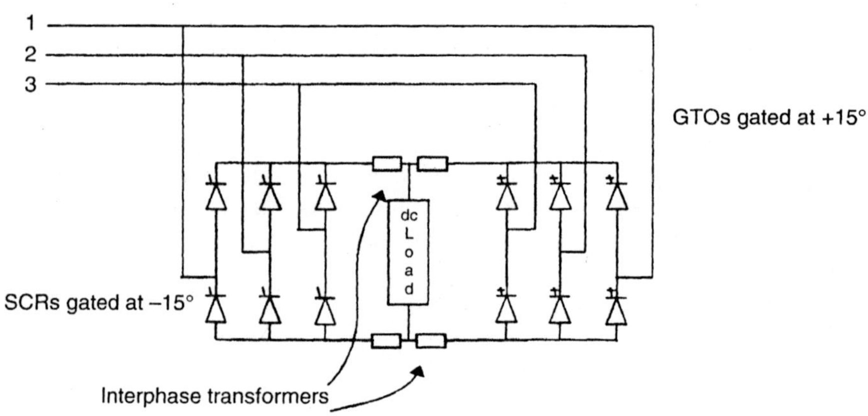

Figure 8-4 Possible electronic phase shift to reduce 5th harmonic.

In a computer simulation the 5th and 7th harmonics were reduced to 5.0% and 8.1%, respectively. A harmonic constant of 202 was obtained for harmonics up to the 25th. Thus, there is little reduction of total voltage distortion with this method. Also, there are two drawbacks with the concept. First, the dc output voltage is reduced to V_{do} cos 15°, that is, 0.96 V_{do}. Second, the use of gated devices introduces higher-frequency harmonics due to the switching actions. No attempt was made to optimize this method, but it is given here to indicate the steps that might lead one to use phase-shifting transformers.

As pointed out previously, a transformer shifts positive- and negative-sequence components in opposite sense. Thus, the transformer provides a mechanism whereby voltages and currents are automatically eliminated in pairs, for example, the 5th and 7th. In contrast, the simple electronic phase shift circuit in Figure 8-4 has fundamental frequency gate control and can only remove one harmonic at a time.

8.4.5 Active Filtering Within the Equipment

The development of variable-frequency drives (VFDs) has led to high-performance microprocessor controls that are capable of functions not available in analog designs. Significant advances have also been made in low-cost pulse-width-modulated (PWM) inverters. These improvements have been coordinated to make a versatile variable-frequency inverter supply. Used with an ac induction motor, a variable-speed performance can be obtained comparable to that of a dc motor. Similar equipment can also be used to control the power conversion process occurring at the converter input.

The motor drive inverter operates from a sensibly constant dc link voltage and is capable of supplying or accepting adjustable frequency power up to the limits of its rating. When accepting power back into the dc link, means must be provided to remove any excess power that might otherwise raise the dc link voltage. The adjustable fundamental output voltage is attained by pulse-width modulation in which the power semiconductors can be turned on and off as required.

The inverter controls usually incorporate a current regulator loop such that the current is under control at all times. It follows the signal demand, which is sinusoidal in steady state, using a PWM process. Motor winding inductance filters the current such that only small perturbations of current occur. The current harmonics are reduced by having a suitably high switching frequency in the pulse-width modulation pattern. For example, 1.2 kHz is easily applied up to a few hundred horsepower. For lower powers, switching frequencies of up to 16 kHz have been used.

The same inverter parts, in conjunction with similar control strategies and faster response, can be used at the power input side of the system to control the amount of power being drawn or supplied to the dc load. In this fashion, the converter operates rather like a synchronous generator. Power flow depends on the inverter frequency/phase and VAR on the converter voltage.

This arrangement requires an appropriate ac line inductance and sufficiently large dc link capacitor to limit dc voltage perturbations. With sufficiently high switching frequency, the ac line current, drawn from the ac source, approaches a sinusoidal waveform. A basic schematic is shown in Figure 8-5 using a 6-pulse GTO converter.

This method of harmonic current control is an emerging technology. A detailed discussion is outside the scope of this book, but a brief review may be helpful. The technique is certainly viable technically, and it is not restricted to low power. A recent article [8] describes a 2000-kW inverter for this application. One design used 6 GTOs. Another design with reduced harmonics used 12 GTOs. The GTOs were rated at 4500 V and 2000 A.

With the 6-GTO converter the lowest-frequency harmonic current was at a frequency of $f_s \pm 4f_o$, where f_s is the switching frequency and f_o is the supply frequency. With a GTO switching frequency of 500 Hz and a source of 50 Hz,

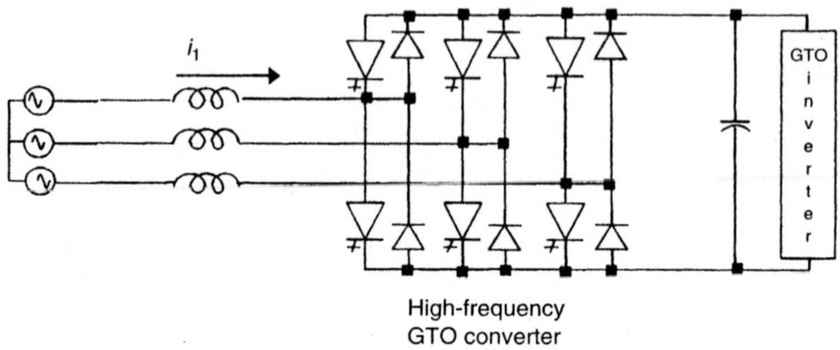

Figure 8-5 PWM input converter controls harmonic currents.

the lowest line current harmonic was 300 Hz with an amplitude of 6%. A harmonic current of about 5.7% was reported at a frequency of $f_s - 2f_o$, that is, 400 Hz. The 12 GTO design had similar harmonic frequencies but at reduced level.

With these input-switching converters, additional passive filtering may be necessary to limit high-frequency harmonic distortion and control the electromagnetic interference (EMI) produced. Because of the converter complexity, the reliability will be lower than that of simple diode rectifiers.

As with the add-on electronic filtering methods, future developments will determine the place of input-switching converters in the order of things for controlling power quality.

8.5 SYSTEM FILTERING METHODS

The analysis in Section 8.2 concluded that deployment of conventional nonlinear loads is limited. One method to increase the tolerable nonlinear load is to use different equipment designs. Another approach is to use power system filters. In this case the power system is evaluated, and appropriate compensation equipment such as tuned filters are added. These techniques are described in the literature and are briefly reviewed here.

As a method, system filters, whether active or passive, have the advantage of being retrofittable but the disadvantage of being possibly only a temporary solution. If the power system changes, for example, if more nonlinear load is added, the design assumptions will also change. In this event, the compensating equipment may become overloaded or ineffective.

8.5.1 Passive Filters

Passive filters can be designed to reduce harmonic voltages and notch effects at particular points in the power system. Each installation is different and the size and placement of the filters varies accordingly. The discussion here highlights some of the design issues and performance characteristics.

Usually the passive filters include different types of parallel paths that present a relatively low impedance to the various harmonics [3, 10, 34]. Harmonic currents flow into this reduced impedance such that the harmonic voltage at that point is reduced. For a power system such as that shown in Figure 8-6, the sources of harmonic current are dispersed throughout the system. These harmonic sources include various types of equipment from various manufacturers, and the question of who is responsible for which filter design may not be obvious. A filter for any particular equipment would be placed near to it to take advantage of source reactance to that point. In some cases, there will be sufficient source impedance at the location at which harmonics must be reduced that a single filter at that location can absorb harmonics from the multiple harmonic sources. This point might be the point of common connection, but in any event the filter must be designed so as not to be overloaded by harmonic currents from other parts of the power system. These can be elaborate calculations.

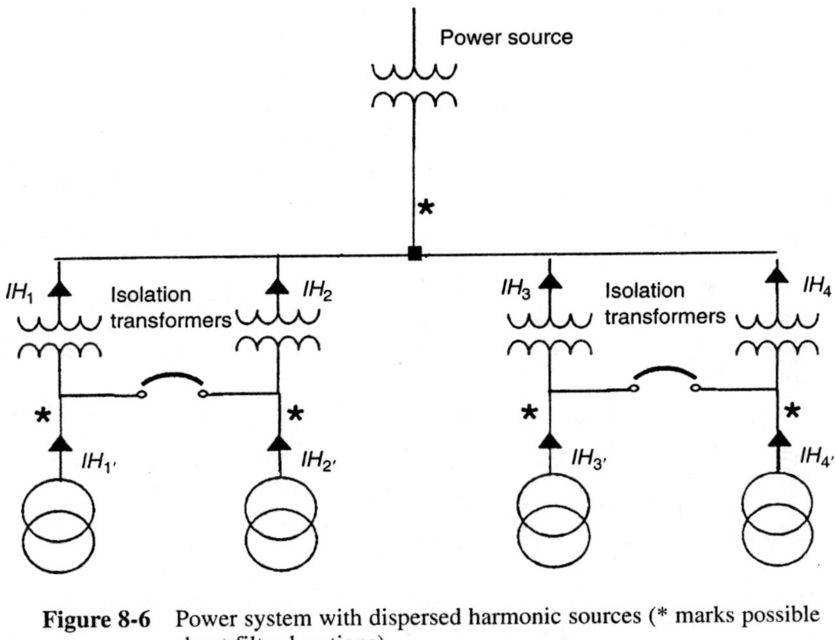

Figure 8-6 Power system with dispersed harmonic sources (* marks possible shunt filter locations).

The power source impedance can often be represented by a simple inductive reactance and the simplest filter approach would connect a single parallel capacitor. However, this may not be satisfactory because of the large capacitor rating required to provide low impedance at the 5th harmonic. The filter capacitor and source inductor will exhibit parallel resonance (high impedance) at a frequency below that at which filtering is effective. This high impedance should not occur at an integer multiple of the supply frequency in case there are harmonics existing at these frequencies.

The size of the filter capacitor can be significantly reduced by connecting an inductor in series with it, and series tuning the combination close to the harmonic frequency. Parallel resonance will still occur but at a higher frequency than would be obtained with an untuned capacitor giving the same attenuation to the harmonic. Parallel resonance effects require evaluation on a case-by-case basis. At frequencies above the tuned frequency, the attenuation of the tuned circuit will be less than that obtained by a simple capacitor.

If a filter is tuned exactly to the harmonic frequency, for example, the 5th harmonic, it will offer a very low impedance at that frequency. This is helpful for filtering, but the tuned circuit will accept 5th harmonic currents from throughout the utility system and may become overloaded. In some cases, filters for the 5th harmonic may be detuned so as to resonate at the 4.7th or 4.8th harmonic [10, 28]. This reduces the filtering effect but also reduces the possibility of overloading.

A general arrangement of multiple parallel-path filters is shown in Figure 8-7. Low-impedance paths for specific harmonics are obtained by series-tuned inductor/capacitor elements connected across the ac supply bus. One or more of the tuned filters can include a resistor across the inductor to attenuate higher-order harmonics. The same capacitance may be feasible in each parallel path, but proportioning the filters to provide low impedance at one frequency without causing unwanted high impedance at another frequency requires detailed design.

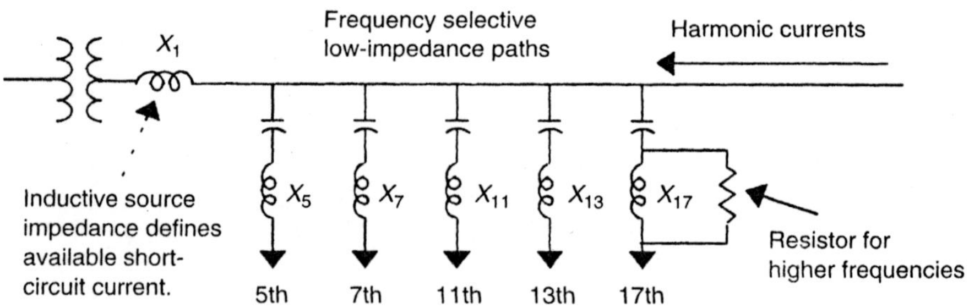

Figure 8-7 Arrangement of multiple parallel path filters using series-tuned harmonic "traps."

Figure 8-7 shows filters for the characteristic harmonics expected from a 6-pulse converter. Other fixed-frequency harmonics can usually be filtered by the same technique. In many practical cases, a single filter path, tuned to the 5th harmonic, may be sufficient.

The overall effect of parallel filters is to provide a low-impedance path for selected harmonics. The actual value of this impedance is affected by the tuning and the inductor Q factor ($\omega L/R$). A value of Q between 20 and 100 is not unusual for filter inductors, but wider variations are feasible. A practical three-phase filter may incorporate a three-phase delta-connected capacitor bank and single-phase inductors.

An idealized impedance plot for a multiple filter and an inductive source is shown in Figure 8-8. At selected frequencies the shunt filters significantly reduce the effective impedance. However, for each low impedance that "traps" a harmonic, a corresponding high impedance is produced at a frequency below the "trap" frequency. For example, in Figure 8-8 the impedance to the 5th harmonic (300 Hz) is reduced by 10 dB, that is, 3.16 times. At 268 Hz the impedance is increased by a factor of 10 times. In practice, system power loads damp the parallel resonance affects and reduce the amplitude of the high-impedance nodes.

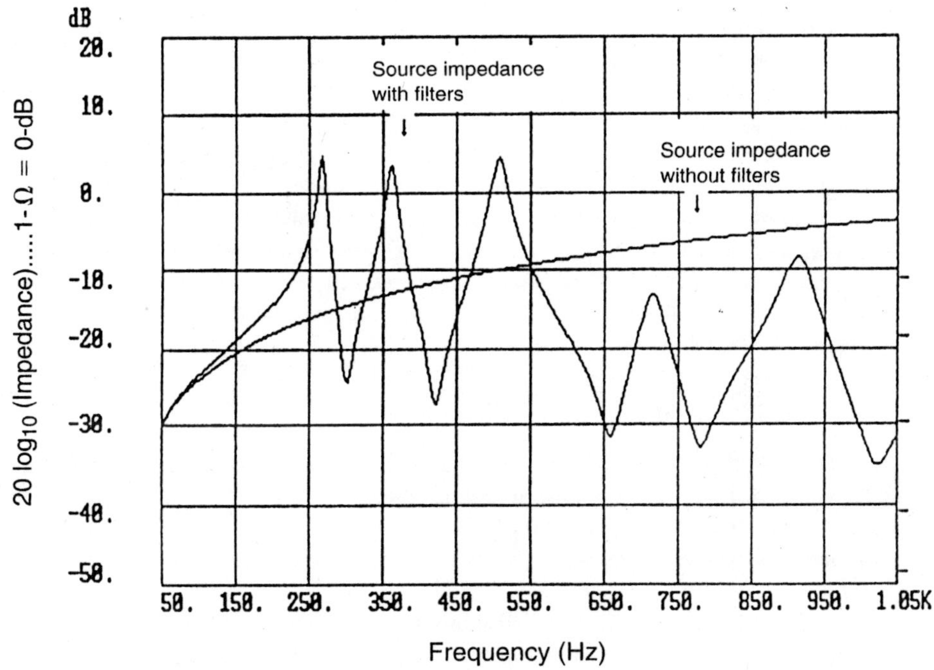

Figure 8-8 Typical impedance "seen" by harmonic currents for the type of filter shown in Figure 8-7. *Note: System power loads will reduce peaks caused by parallel resonance.*

The final filter design must allow for component tolerances and system variations. For example, initial inductor and capacitor values typically range within ±5% of nominal; also, changes occur due to temperature and other operating conditions. Tappings on the inductors can provide for field adjustments. Sometimes the utility will switch power factor correction capacitors in and out of service on a daily basis. This modifies the assumption of a simple inductive reactance for the utility source impedance. Also, the parallel filters provide leading kVAR that may modify the amount of power factor correction required. These changes are conveniently addressed by computer simulations, once the system data has been obtained.

The tuned-filter technique can be applied to other types of power electronics equipment; however, as noted previously, it may not provide a permanent solution.

Passive filters are much more difficult to use in conjunction with auxiliary generators for two reasons. First, the generators cannot normally support more than about 20% leading kVAR because armature reaction may cause overexcitation and voltage regulator instability. This would especially be a concern when the power converter is on light load [15, 32]. Second, frequency variations expected with an auxiliary generator are much greater than those of the utility; therefore, filter design is complicated.

Passive filters are widely used in conjunction with utility-type static VAR compensators and ac electric arc furnaces with megawatt ratings [16]. In this type of application, the major source of harmonic disturbance is well known, and the probability of system changes affecting the filter performance is small. (dc-type arc furnaces could be supplied from multipulse converters, as a means for controlling harmonics; however, that discourse is outside the scope of the present discussion.)

8.5.2 Add-On Active Filters

In this method an additional power electronic converter is used to supply the power source line with the harmonic currents required by the nonlinear load. In essence, the filter is a power amplifier and must have adequate bandwidth to compensate for the harmonic currents required by the electronic equipment, at least up to the 25th harmonic.

Technically, this method is undoubtedly very effective. The main drawback lies in its cost, which, with development, is expected to be comparable to an inverter of similar rating. In contrast with typical motor drive inverters, which operate from a stable dc link voltage, the active filter is exposed to voltage stresses caused by normal and fault conditions in the power system. This puts additional demands upon the semiconductor switching devices. Hybrid arrangements of active and passive components are also feasible [30].

At the time of writing, these products continue to be developed. Their eventual place in the order of ways to correct for system harmonics will depend on economic issues.

8.6 PHASE STAGGERING

This technique for reduction of harmonics may be helpful when there are multiple loads that can be fed from different phase-shifting transformers. The method is really a combination of equipment and system variations.

Reduction of plant harmonic currents is obtained by feeding separate power electronics equipment through appropriate 1:1 phase-shifting transformers. Conventional double-wound, delta/wye isolation transformers have traditionally been used to give 12-pulse system performance and reduce 5th, 7th, 17th, and 19th harmonic currents.

Delta/wye transformers are routinely available; however, greatly improved efficiency and lower cost are attainable with auto-connected polygon or differentially rated designs. The polygon transformer is discussed in detail in Chapter 5, Section 5.1.

With auto-connected phase-shift methods, the angle of phase shift is not limited to the traditional delta/wye phase shift of 30°. Using appropriate phase shifts, other harmonic currents such as 11th and 13th can also be reduced.

The essence of phase staggering is illustrated in Figure 8-9, which shows two 6-pulse converters fed from the same source. One converter bridge is fed directly from the source. The other converter is fed from an auto-connected polygon transformer with a 30° phase shift.

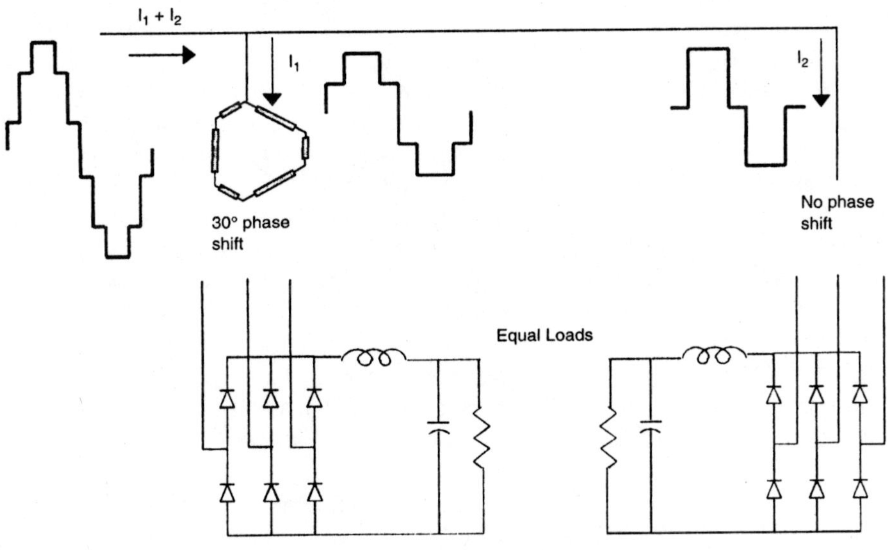

Figure 8-9 Two three-phase bridges "staggered" to draw 12-pulse current.

Referring to Figure 8-9, each converter is assumed to have the same load and to draw fundamental and harmonic currents of the same magnitude.

The waveshapes are different because the phase angle of the $6(2k-1)\pm1$ harmonic currents are different—by 180°—as it happens. Thus, the 5th, 7th, 17th, and 19th harmonics of one converter are supplied by the harmonics of the other. The net result is that there are no 5th and 7th harmonics of current drawn from the source.

This method of harmonic elimination is most effective when the converters are identical and equally loaded. Under varying load conditions, an idealized worst case occurs when only one converter is operating and it has full load. In this event, the currents are half what they would be with two converters fed directly (no phase shift) from the line.

Slightly worse results may occur at full load when, due to practical variations in individual converters, there are residual harmonics due to imperfect cancellation. In Chapter 7, residuals of 20% were recommended to be used for conservative initial calculations. In this regard, the arrangement shown in Figure 8-9 would give better symmetry and hence better cancellation if each converter was fed through a 15° phase-shift polygon, one with +15°, the other with −15°.

Phase staggering to simulate 12-pulse operation is not as effective as a 12-pulse converter in which residual harmonics can, in principle, be controlled. Nevertheless, it is a simple method which can significantly raise the tolerable amount of nonlinear load on a system. If we refer back to the calculations in Section 8.2.4, we would expect 12-pulse to about double the permissible loading, from 29% to 58%, if only the 5th harmonic of current is considered. Unfortunately, the 11th harmonic of current, which is not affected by 30° phase staggering, now becomes the limiting factor, and the load is restricted to about 40%. Despite this, we have moved much closer to meeting the harmonic current requirements by the phase-staggering method.

Figure 8-10 illustrates another method of subdividing the load. This time there are three equal rating converters fed in phase-staggered fashion. One is fed directly from the line, the others are fed with phase shifts of +20 and −20°. Waveforms for this are shown in Figure 8-11.

The resultant harmonic current at full load includes harmonics of the form $(18k \pm 1)$. Neglecting residuals, worst case applies under two conditions, namely, when one or two converters are operating at full load. In this event, the 5th, 7th, 11th, and 13th harmonics are one-third that of a single three-phase bridge of full rating.

In this example of phase staggering, the 17th harmonic becomes the critical factor. IEEE Std 519-1992 limits the 17th harmonic of current to 2.5% for I_{sc}/I_L greater than 20 and less than 50. From Table 7-1, a value of 2.74% is calculated for the 17th harmonic of current when the ac line reactance is 5%.

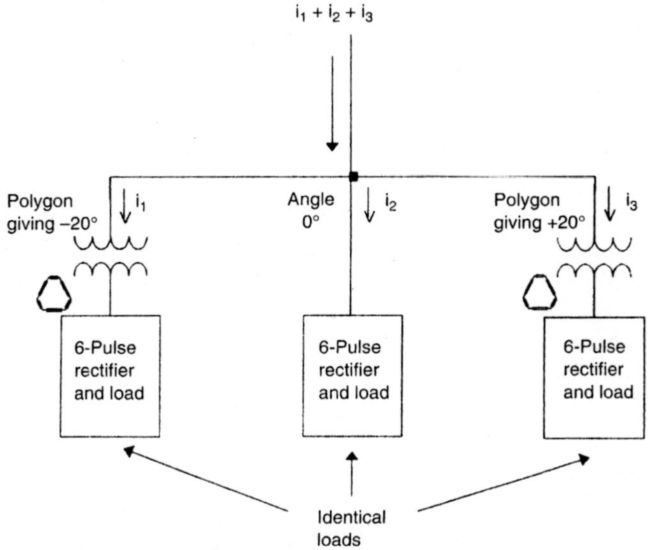

Figure 8-10 Three three-phase bridges "staggered" to draw 18-pulse current.

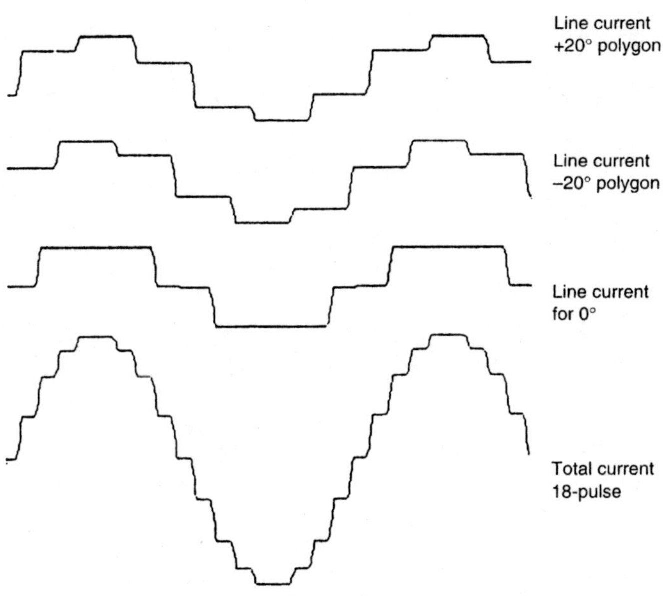

Figure 8-11 Waveforms for three phase-staggered converters as in Figure 8-10.

If we look again at the calculations in Section 8.2.4 and assume the transformer is fully loaded, case (b), then the ratio I_{sc}/I_L remains the same. Solving as before, about 90% of the utility transformer load can now be three-phase rectifier bridges.

If we assume only nonlinear loading as in case (a), then the short circuit ratio is changed to less than 20 and a lower value of 17th harmonic of current (1.5%) is mandated from Table 1-2.

Taking all this into account, it is feasible for the utility transformer to be loaded with about 85% nonlinear three-phase bridge rectifier loads when they are balanced as shown in Figure 8-10. Clearly, it is unlikely that the total load can be grouped exactly as shown in Figure 8-10. Nevertheless this technique appears useful for many practical systems. It is effective and efficient.

The "staggering" method can be further extended to use four equal loads and thence provide 24-pulse characteristics at full load. Various grouping methods can be used in which the four converters are fed with power supplies spanning a range of 45° in 15° increments. The method shown in Figure 8-12 requires only three phase-shifting transformers with a total equivalent rating totaling less than 25% of the total power load.

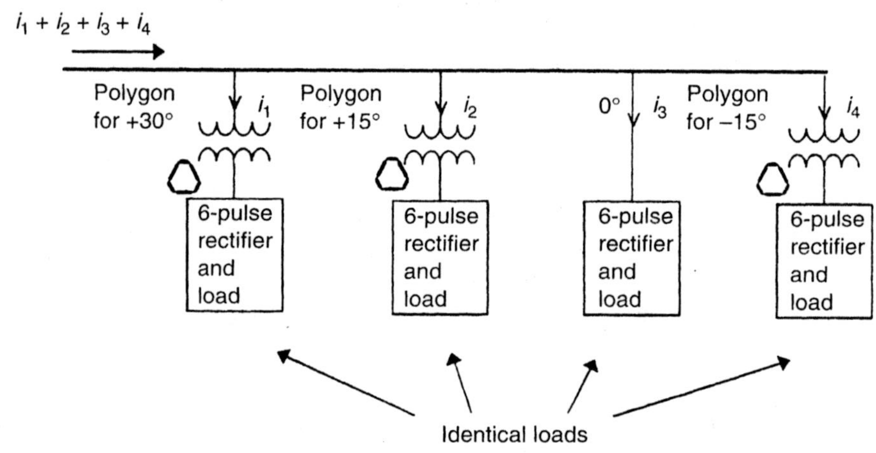

Figure 8-12 Four staggered converters draw 24-pulse current.

Figure 8-13 is a vector diagram representing the phase position of harmonics in each converter. It can be used to determine the residual harmonics when various converters are unloaded.

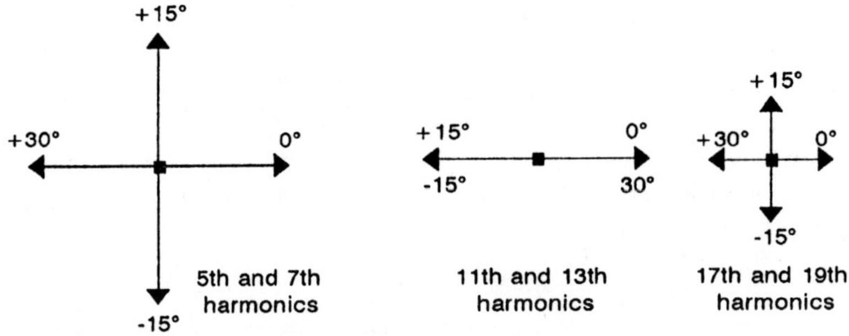

Figure 8-13 Harmonic vector phase relationships for converters staggered by +30°, +15°, 0°, and −15°.

With one 6-pulse unit out of service, harmonic currents in the supply will include the harmonics normally supplied by that unit. These include the 5th, 7th, 11th, 13th, 17th, and 19th harmonics. These are present in an amount that is effectively 25% of those that would be present if the complete load was fed by a 6-pulse converter. Thus, the 5th harmonic is about 5% of the four-converter rating.

If two units are out of service, the harmonics will depend upon which two units are inactive. For example, if the +15° and −15° degree units are unloaded, the other two units will provide 12-pulse performance at half-rated current. If the units at 0° and −15° are out of service, the power source will include the 5th, 7th, 17th, and 19th harmonics of current, which are $\sqrt{2}$ times the values of a single 6-pulse unit; however, there are no 11th or 13th harmonics.

Separation of the total load into four equal sections helps reduce the system harmonic currents and may be desirable depending upon the individual system load configuration. The method has been used extensively in rectifiers for aluminum smelting.

So far we have considered staggering the supplies to multiple 6-pulse rectifiers fed from the utility bus. Similar calculations can be made for auxiliary generator sources; however, these have only about one-third the short-circuit capability, and generally I_{sc}/I_L will be <20. Because of this we must meet exacting requirements for harmonic currents as well as voltage. Several practical examples have shown that the need for less than 5% total harmonic voltage distortion is likely to be the limiting criterion for auxiliary generator systems.

Individual phase-staggering cases require separate analyses. As a guide, Table 8-2 can be used to anticipate the voltage harmonic constant. Detailed calculations would then follow.

TABLE 8-2.

APPROXIMATE RELATIVE HARMONIC VOLTAGE CONSTANTS FOR 6-PULSE RECTIFIERS ($X_{pct} \approx 3.0$)*		
Circuit Arrangement	**Capacitor dc Filter**	**Inductor dc Filter**
Direct connection	100%	76%
0° and 30° phase stagger	50%	38%
+20° 0° −20° phase stagger	34%	26%

*This data is useful only for initial estimates with phase staggering. It assumes balanced loading.

8.6.1 Phase-Staggering Example

Assume that, with a number of inductor-input filter converters installed, initial calculations indicate a voltage distortion of 7.5% at full load. Also assume that the load could be split up into two parts in the ratio of 55% to 45%. To reduce voltage distortion to 5.0%, we need to reduce the harmonic constant by a factor of 5.0/7.5, that is, 0.67. From Table 8-2 a single 30° phase-shifting transformer feeding 50% of the load would reduce distortion by a factor of 38/76, that is, 0.5. Results will probably be satisfactory with only 45% load, but detailed calculations are recommended to verify this.

Figures 8-14 and 8-15 illustrate the fifth harmonic current when two converter loads of equal rating are staggered by means of a 30° phase-shifting transformer. For best results the converters should incorporate a dc filter inductor.

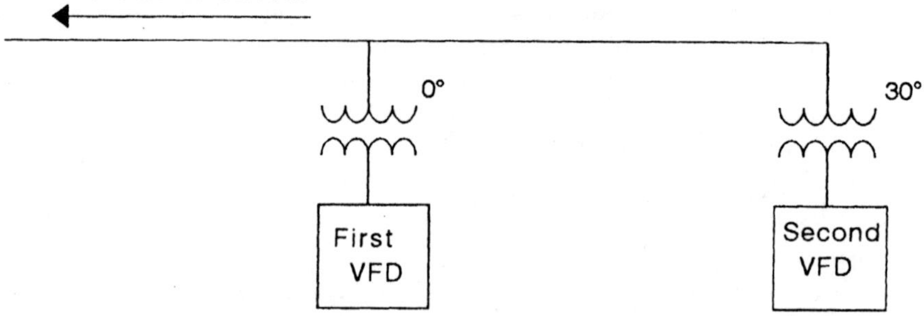

Figure 8-14 A schematic of two equal rating VFDs with phase staggering.

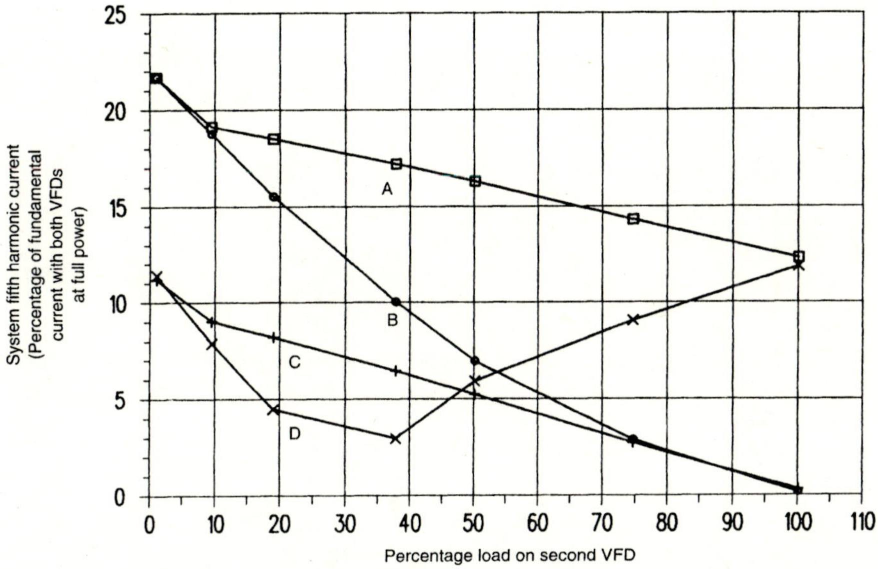

Figure 8-15 Calculations from the schematic shown in Figure 8-14. *Note: each VFD has 2.2% ac line reactance; source is type 1 power.*

Key for Figure 8-15		
Curve	**First VFD 100% fixed load @ 0°**	**Second VFD variable load @ 30°**
A	Capacitor dc filter	Inductor dc filter
B	Capacitor dc filter	Capacitor dc filter
C	Inductor dc filter	Inductor dc filter
D	Inductor dc filter	Capacitor dc filter

8.6.2 Summary of Phase Staggering

The use of phase-staggered converters appears to offer a cost-effective means for controlling distortion of voltage and current in some installations. Initial selection of the load to be staggered is straightforward, but calculation is necessary to confirm the performance. Where appropriate loading exists, phase staggering may offer a method to reduce voltage and current distortion by more than 2:1.

Harmonic current limitations will often require much greater reduction than is attainable by staggering means. When this method is inadequate, other means to filter within the power electronics equipment can be applied.

8.7 MEASUREMENT OF SYSTEM VOLTAGE DISTORTION AND HARMONIC CURRENTS

8.7.1 The Position of Measurements

Measurements of voltage and current should be made at the point of common coupling (PCC), or at any point mutually agreeable to the consumer and the utility that enables the harmonic performance at the point of interest to be determined. Monitoring equipment for individual loads should be at a convenient point close to the load.

8.7.2 Measurement of Voltage Distortion

IEEE Std 519–1992 aims to limit individual harmonic voltages to less than 3.0% by means of specifying individual harmonic currents, but it does not require individual voltage harmonic measurements. It defines total harmonic voltage distortion, which can be measured with readily available equipment. If system changes must be made to bring the performance into compliance, then measurement of individual harmonic voltages may be very helpful. Much of the commercially available measuring equipment has this capability. Measurement of total voltage distortion is seen to be relatively straightforward.

Voltage notching is not quite so straightforward but can be determined with the help of an oscilloscope. As described in Chapter 1, the anticipated notch area and location within the cycle will vary depending upon the specific converter topology.

8.7.3 Measurement of Harmonic Currents

Total individual harmonic currents at the PCC can be determined quite simply using one of the commercially available instruments. With knowledge of the total fundamental current I_L and short-circuit current I_{sc} at the PCC, conformance with IEEE Std 519-1992 can be determined in a straightforward manner. Determination of the harmonic current "generated" by individual nonlinear loads is not so straightforward. This is because preexisting harmonic voltages will affect the harmonic current drawn by linear and nonlinear loads. This has already been emphasized in Chapter 2 where the performance of some multipulse systems was seen to degrade in the presence of 2.5% of 5th harmonic voltage.

Topologies to overcome this sensitivity are given in Chapter 9, but there remains a residual effect. For example, a linear resistor load fed with 2.5% of 5th harmonic voltage will draw 2.5% of 5th harmonic current. Nevertheless, it would be unreasonable to conclude that it "generates" 5th harmonic current. If it is necessary to reduce harmonic currents at the PCC and a search is made for the equipments "generating" harmonic currents, the effects of system harmonic voltages must be considered. Whether they are significant or not depends upon the system. As noted in Chapter 7, the total harmonics associated with 6-pulse converters are so large that they are not much affected by practical levels of preexisting harmonic voltages. It is in the nearly linear loads, such as 18-pulse low-harmonic converters, that the influence of preexisting line voltage harmonics will have the largest percentage effect; however, the resulting harmonics will still be small, as shown in Figure 10-9.

Chapter 9

Multipulse Circuit Performance

9.1 INTRODUCTION

Six multipulse circuits are selected and reviewed in this chapter. They appear to be especially suitable for 480-V application, such as may be required in the adjustable-frequency drives arena.

Two new circuit concepts using midpoint connections are presented and analyzed. They may require refinements for the marketplace, but they offer excellent robust performance for a wide range of applications. All circuits have performed well in computer simulations that incorporate practical parameter variations.

These circuits have, to some extent, been the subject of classical analysis and discussion during evaluation of phase-shifting transformers in Chapters 4 and 5. Highlighting them here facilitates comparisons and prepares the way for computer simulation. Particularly, these circuits are used to highlight the affects that source parameters may have. For convenience in discussion, the circuits are given multipulse connection (MC) reference numbers as follows.

MC-101. 12-pulse with conventional delta–delta/wye double-wound transformer.

MC-102. 12-pulse with two identical auto-connected $\pm 15^{\circ}$ phase-shifting transformers and two identical interphase transformers.

MC-102A. Variation of MC-102 in which a single interphase transformer is used in conjunction with a zero-sequence blocking transformer.

MC-103. 12-pulse with hybrid delta/wye transformer.

MC-104. 18-pulse with differential fork step-down transformer.

MC-105. 12-pulse with differential fork step-down transformer.

The schematics and design information for these circuits are given in the appropriate figure. The design formulas given are idealized. They relate to conditions in which the dc current I_d has no ripple and in which the effects of ac line reactance are omitted. Since neither of these conditions will apply in practical designs, it is necessary to review what changes can be expected when these idealized conditions are not met.

Before establishing the range of practical features to be included in computer simulations, some discussion of commutation in multipulse converters is desirable.

9.2 COMMUTATION EFFECTS

The filtering effect of ac line reactance, which slows the rate of change of current at commutation, is very significant in "inherent" higher-pulse-number, midpoint-connection converters. The term "inherent" refers to multipulse connections that cannot be broken down into conventional parallel or series combinations of 3-pulse circuits.

Computer results show the effects of ac line reactance quite clearly, but a qualitative explanation provides further insight. In the "inherent" multipulse midpoint connections, the device conduction period is reduced. For example, it is ideally only 40° in an 18-pulse arrangement with 9-pulse commutating group. Because of this reduced conduction interval, the ac line reactance has a larger total effect than on a pulse that is of 120° duration, as in a 3-pulse commutating group, for example.

Commutation groups refer to a combination of devices in which the load current transfers from one device to another in a cyclic manner. Comparisons are simplified by assuming that commutation is unaffected by other parts of the circuit. This is usually true in multipulse connections. When necessary, the computer can easily accommodate simultaneous commutations.

The response of multipulse converters to ac line reactance depends upon the number of devices in the commutating group.

Figure 9-1 illustrates a general commutating group. The "incoming" device (current rising) and "outgoing" device (current falling) follow cosine shape waves [2]. The point to be made here is that the current is dependent upon the voltage existing between the devices. In general, the commutation voltage is $\sqrt{2}V_N$ 2 sin $\Phi/2$, where Φ is the angular difference between voltages in the commutation group.

In a 3-pulse commutating group, such as occurs in a three-phase bridge converter, Φ is 120°, and the commutation voltage amplitude is $\sqrt{6}V_N$. For the 18-pulse case, the vectors are at an angle of 40°, and the commutating voltage is only $0.967V_N$. In this 9-pulse group, there is a further reduction in commutating

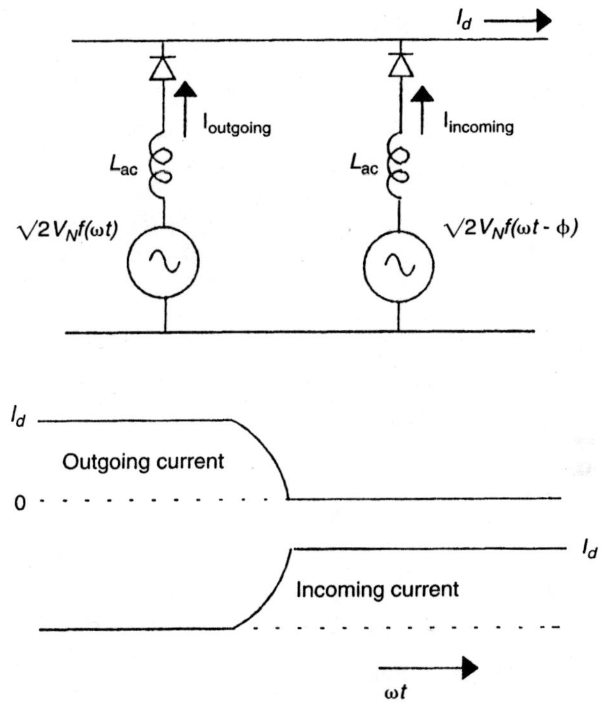

Figure 9-1 Transfer of current from one device to the next in a commutating group.

voltage because the required line-to-neutral voltage for a given dc output is less. The net result is that the amplitude of the commutating voltage is only $0.85V_N$. Thus, for a given circuit inductance, the rate of change of current in a 9-pulse group is reduced by a factor of at least 2.8 times. The effective impedance of L_{ac} is magnified when it is referred to the transformer primary source voltage. This is why notching effects are of little practical concern with this type of 18-pulse converter.

These effects also mean that the voltage drop associated with ac line reactance is increased. A computer simulation showed that the 17th harmonic of current could be reduced to less than 1.5% with a full-load voltage drop of only 5%. The ac line reactance to achieve this is only 1.44%.

An appropriate terminology is needed to establish the effects of ac line reactance. Also for these purposes it is assumed that commutations do not occur simultaneously. This will normally be the case; however, the computer simulations easily handle different situations.

9.3 DEFINITION OF AC LINE REACTANCE

The technique of using percentages has been used very effectively in conjunction with power transformers. We will use the same method for defining ac line reactance. The percentage reactance is defined in relation to the fundamental component of current I_1 flowing through the circuit and the value of short-circuit current I_{sc}. This short-circuit current depends upon the open-circuit voltage V_{oc} available at the point and upon the ac line reactance ωL_{ac}. Thus,

$$\text{percentage ac line reactance} = X_{pct} = 100\frac{I_1}{I_{sc}} \tag{9.1}$$

For these calculations, the circuit resistance is neglected and the inductance is considered linear. In this case the impedance is simply ωL_{ac}. Using per phase values, we get

$$I_{sc} = \frac{V_{oc}}{\omega L_{ac}} \tag{9.2}$$

9.3.1 Example of Percent Reactance Calculation

In this example we use an 18-pulse converter as in Figure 9-9 with the following parameters:

$$V_{do} = 676\text{ V};\ I_d = 200\text{ A};\ L_{ac} = 150\ \mu\text{H}$$

Calculation. From Figure 9-9, system $V_N = V_{do}/2.44 = 277$ V. From equation (5.31), voltage to bridge $= 0.879\ V_N = 243.5$ V, and $\omega L_{ac} = 0.05655\ \Omega$. Thus, $I_{sc} = 4306$ A.

From Example 4 of the appendix,

$$I_1 = \left(\frac{4}{\pi\sqrt{2}}\right)\sin\left(\frac{\pi}{9}\right)I_d = 0.308 I_d$$

Thus,

$$X_{pct} = 100\frac{0.308 I_d}{4306} = 1.43\%$$

More than just commutation effects will be addressed in practical computer simulations for the selected circuits. Nevertheless, some standardization of design conditions is necessary to provide a basis for comparison of practical designs. In this regard, type 1 and type 2 power systems were defined in Chapter 2. Also, it has been recommended that when very-low-harmonic currents are required, the ratio of dc circuit inductance to resistance ($\omega L_{dc}/R_{dc}$) be appropriately defined at full load.

In this chapter, unless noted otherwise, the practical operating conditions are those highlighted in the paragraphs that follow.

- Power source is type SP2 with 1% negative-sequence voltage and 2.5% preexisting 5th harmonic voltage.
- The dc filter circuit incorporates filter inductance such that $\omega L_{dc}/R_{dc}$ = 0.168 at full load.
- Where used, the effective transformer leakage reactance or additional ac line reactance is 3%.
- Source ac line reactance is balanced and less than 1%.
- Inductors have a Q factor of 15 or larger at the fundamental operating frequency of 60 Hz.
- Interphase transformers have a peak-to-peak magnetizing current less than 20% of the total dc load current.
- Transformer turns ratios are the ideal values.
- A large dc filter capacitor is connected across the resistor load to keep that part of the dc voltage practically ripple-free.

With this background complete, we can now proceed to review practical features of the selected circuits.

9.4 TWELVE-PULSE WITH CONVENTIONAL DELTA/WYE DOUBLE-WOUND TRANSFORMER, CIRCUIT MC-101

The schematic and idealized design information are given in Figure 9-2. Computer-generated results are in Figure 9-3.

9.4.1 Discussion of Circuit MC-101

Turns Ratios. The actual turns ratio of $\sqrt{3}$ is an irrational number that we can only approximate in a practical design. The major concern in this circuit is in achieving balance between the paralleled converters. Any voltage difference in the V_{do} of each converter, such as will be caused by an incorrect turns ratio, will cause a current unbalance. This, in turn, prevents cancellation of the 5th, 7th, 17th, 19th, and so on, harmonics.

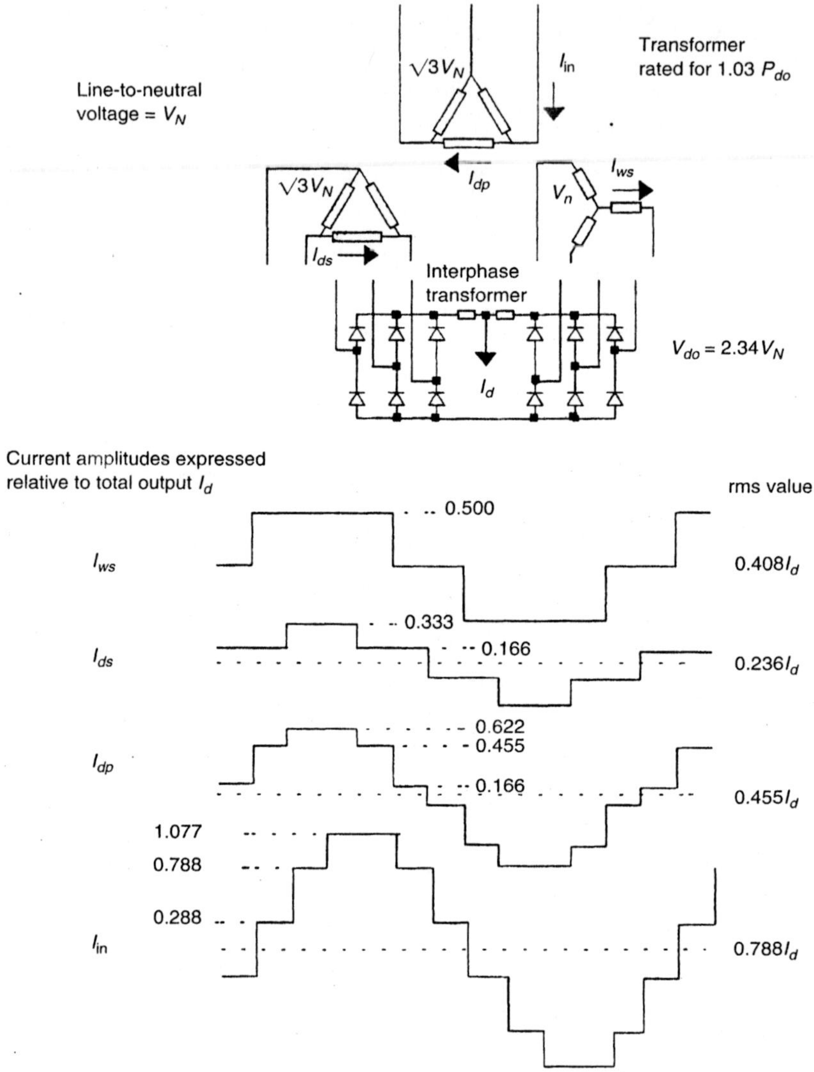

Figure 9-2 Twelve-pulse with conventional delta–delta/wye double-wound transformer, MC-101.

Practical turns ratios within almost 1% of $\sqrt{3}$ can be obtained using the ratios shown in Table 9-1.

Computer results shown in Figure 9-3 highlight the significant effects of 2.5% preexisting 5th harmonic voltage in the source.

TABLE 9-1.

PRACTICAL TURNS RATIOS FOR CIRCUIT MC-101				
4:7	7:12	11:19	15:26	19:33

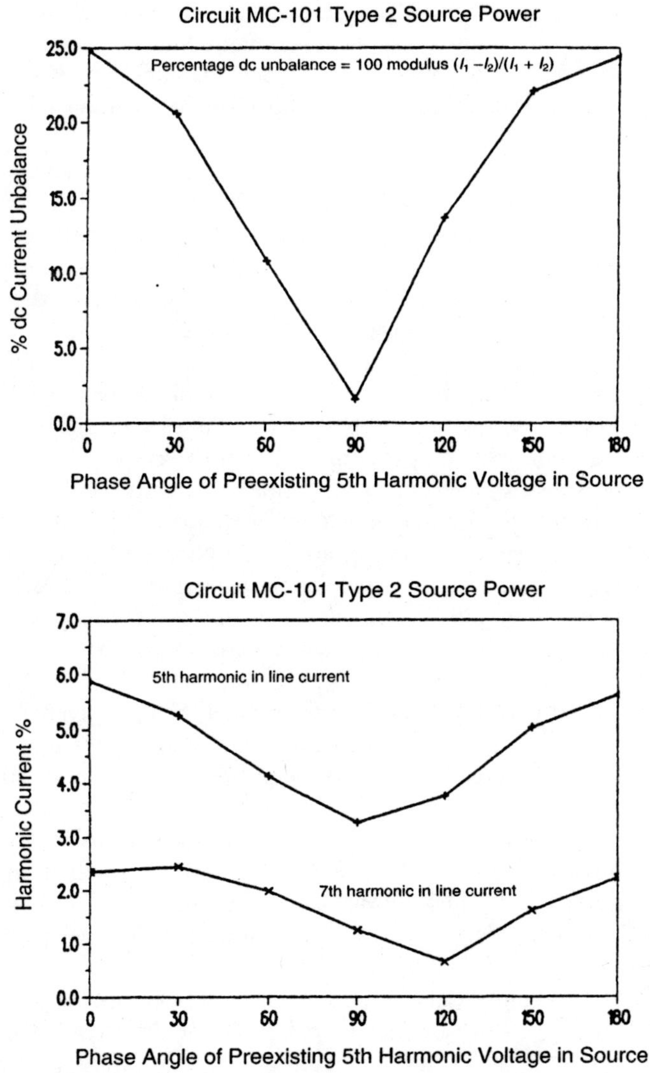

Figure 9-3 Computer results for the 12-pulse circuit, MC-101.

This causes dc load unbalance, which complicates interphase transformer design, and introduces an uncharacteristic 5th harmonic of ac line current. These results are in agreement with the analysis in Chapter 2.

Negative-sequence voltage in the ac source produces 3rd harmonic line current in this 12-pulse circuit in a similar manner to that of a 6-pulse circuit. (See Section 7.2.3.)

Transformer leakage or additional reactance in conjunction with circuit resistance can provide series impedance to help limit the current unbalance. However, with practical amounts of impedance, the unbalance is still significant. Because unbalance severely complicates the design of interphase transformers, it may be desirable to correct for unbalance.

When thyristors are used, balance is obtained by appropriate phase control. In other circuits introduction of compensation by means of high-frequency, low-power dc-to-dc converters may be appropriate. Harmonic blocking transformers, such as that shown in Figure 6-16, may also be considered. As demonstrated previously, the phase angle of preexisting harmonics can rarely be defined in practice. Thus, designs should anticipate worst case conditions.

The interphase transformer helps to keep each converter acting independently of the others. However, this is only applicable above a certain load current. The load must be such that the interphase transformer magnetizing current can flow and the converter output current is continuous.

When the converter load currents become discontinuous, the device conduction is reduced, and the dc output voltage increases. The two 6-pulse bridges compete for the dc output, and each, in turn, supplies a 30° conduction interval. The output voltage has 12-pulse ripple and a corresponding increase of 3.6% in the amplitude of the dc output. This is less than the 4.7% increase due to peak smoothing caused by the load capacitor at light loads. Therefore, open-circuit voltage levels are not excessive.

The computer-calculated results for circuit MC 101, given in Figure 9-3, show that, although transformer turns ratios and impedances may be carefully designed, the power source can still have significant effects on the converter performance.

In the computer simulation, an ac reactance of 3% was included in series with each converter bridge. With a 2.5% 5th harmonic voltage in the source, the worst case of full-load conditions caused one converter to draw 127-A and the other, 73-A. This dc unbalance causes a dc bias in the interphase transformer flux.

The final interphase transformer design must accommodate both rms currents, which affect sizing of the copper, and unbalance and ripple current, which affect sizing of the iron. The resulting interphase transformer is, in practice, much larger than one might expect from an idealized analysis.

9.5 TWELVE-PULSE WITH ±15° AUTO-CONNECTED POLYGON TRANSFORMER, CIRCUIT MC-102

This arrangement requires a total of only 25% for the transformer kVA and provides excellent symmetry. The symmetry helps to balance the individual bridge outputs under some conditions. See Figure 9-4 for schematic and idealized

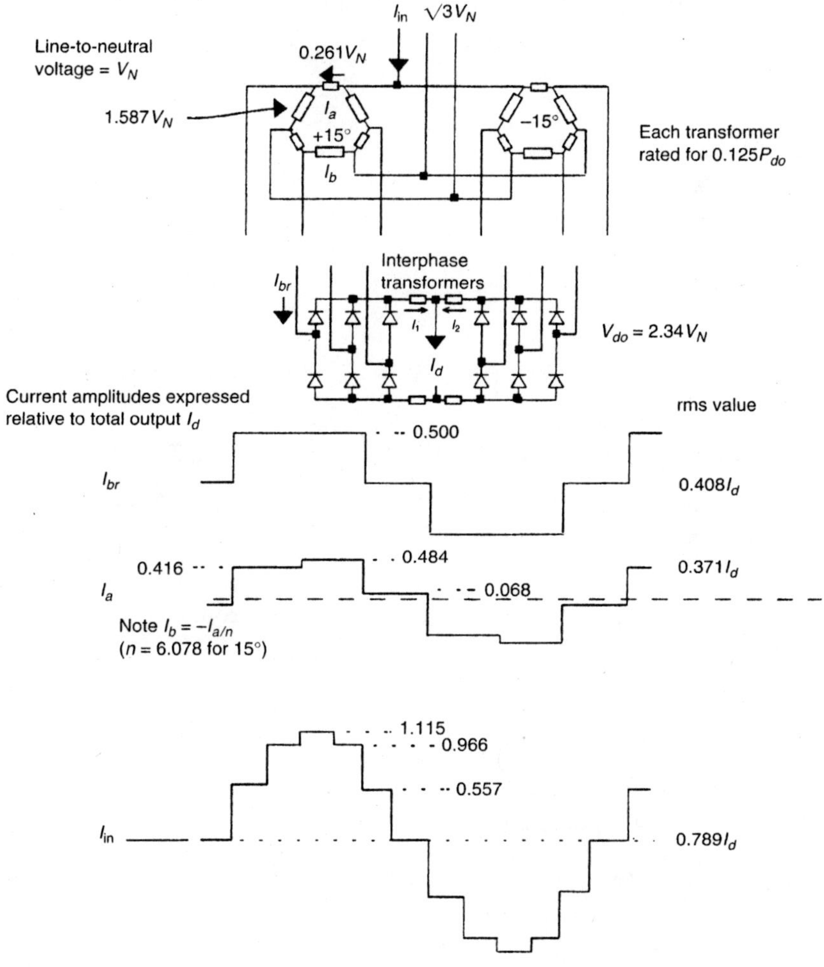

Figure 9-4 Twelve-pulse with two identical auto-connected, phase-shifting transformers and two identical interphase transformers MC-102.

design information and Figure 9-5 for computer-generated results. However, we will see once again that without additional compensation, preexisting harmonic voltages wreak havoc with converter balance. The two separate transformers could be replaced by a single transformer, as shown in Figure 5-6. In this case the dc output voltage will be about 3.5% greater.

Turns Ratios. For a phase shift of 15°, the turns ratio required is 6.078 : 1 (from sin(60°–7.5°)/sin 7.5°). This is a very convenient ratio approximated by 6:1, which yields 15.178°. Thus, a wide range of turns ratio is available. Unlike

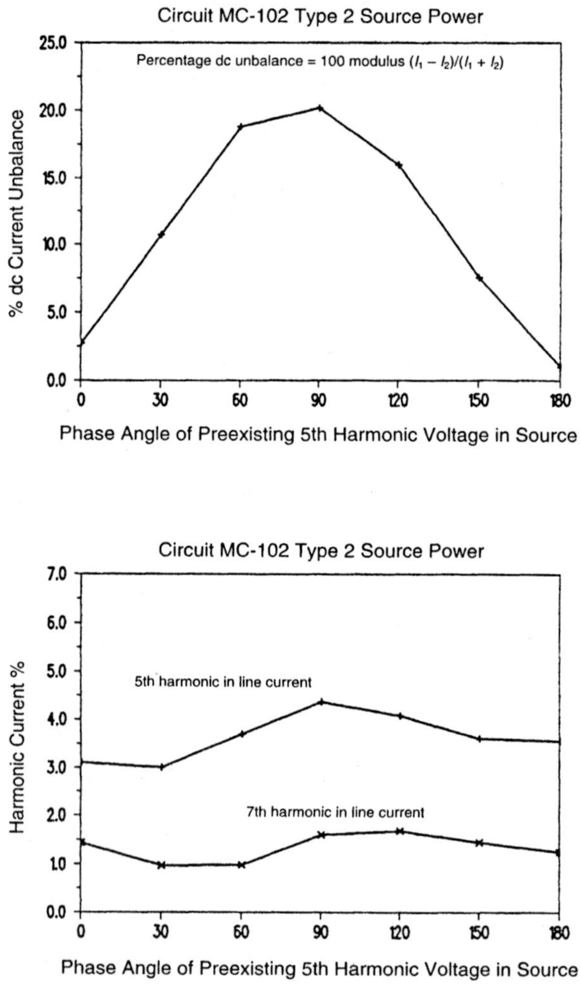

Figure 9-5 Computer results for 12-pulse circuit, MC-102.

the delta/wye arrangement in which the voltage ratios could never be met exactly, the polygons give a 1:1 voltage ratio on open circuit.

Two interphase transformers are used in circuit MC-102. Their rating is considerably larger than the single unit needed in circuit MC-101. At light loads when the current is insufficient to magnetize the interphase transformers, the dc output voltage changes to that of a 6-pulse design. In Figure 6-8, it is shown that without interphase transformers, the effective line-to-line voltage $V_{L\text{-}Leq}$ is the difference of two phase vectors at 150°. The net result is an increase of 11.5% in the inherent open-circuit voltage V_{do}. This voltage increase represents a practical limitation that must be considered in the design of converters to power dc to ac inverter systems.

This limitation can be mitigated by using a zero-sequence blocking transformer in conjunction with a single interphase transformer. It is shown in Figure 9-6 as circuit MC-102A.

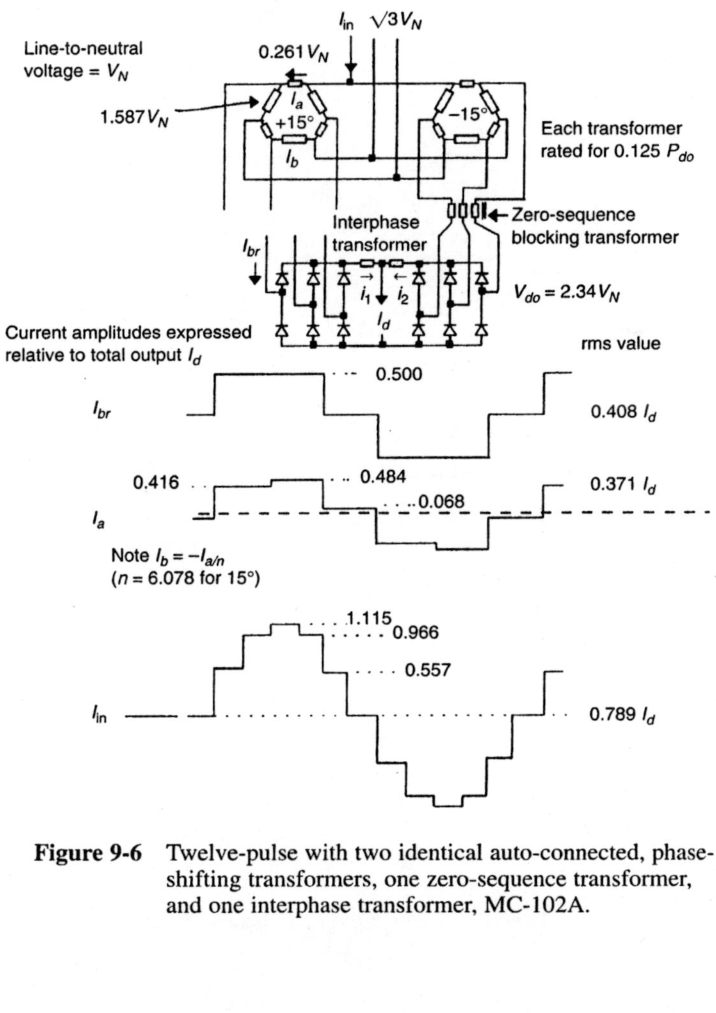

Figure 9-6 Twelve-pulse with two identical auto-connected, phase-shifting transformers, one zero-sequence transformer, and one interphase transformer, MC-102A.

9.6 TWELVE-PULSE WITH HYBRID DELTA/WYE TRANSFORMER, CIRCUIT MC-103

The schematic is shown in Figure 9-7, with basic computer results in Figure 9-8. This connection requires only a simple modification to a standard delta/wye configuration to provide power for two converters in series. For optimum size of the transformer, it should be designed for the application and take advantage of

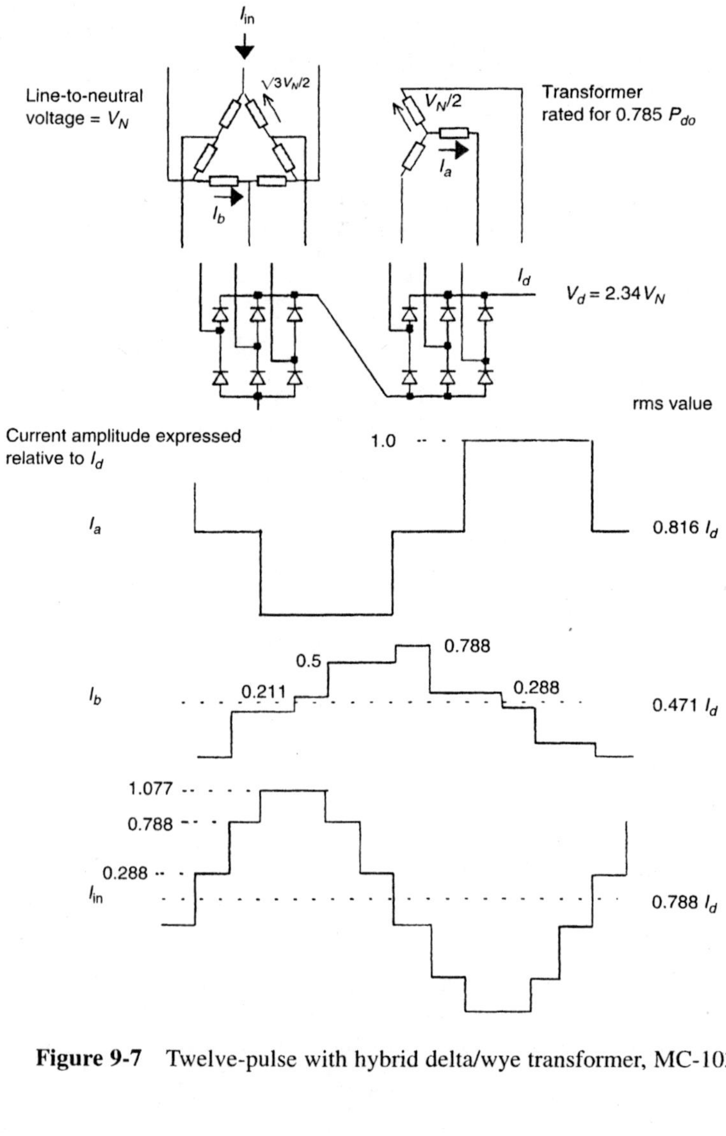

Figure 9-7 Twelve-pulse with hybrid delta/wye transformer, MC-103.

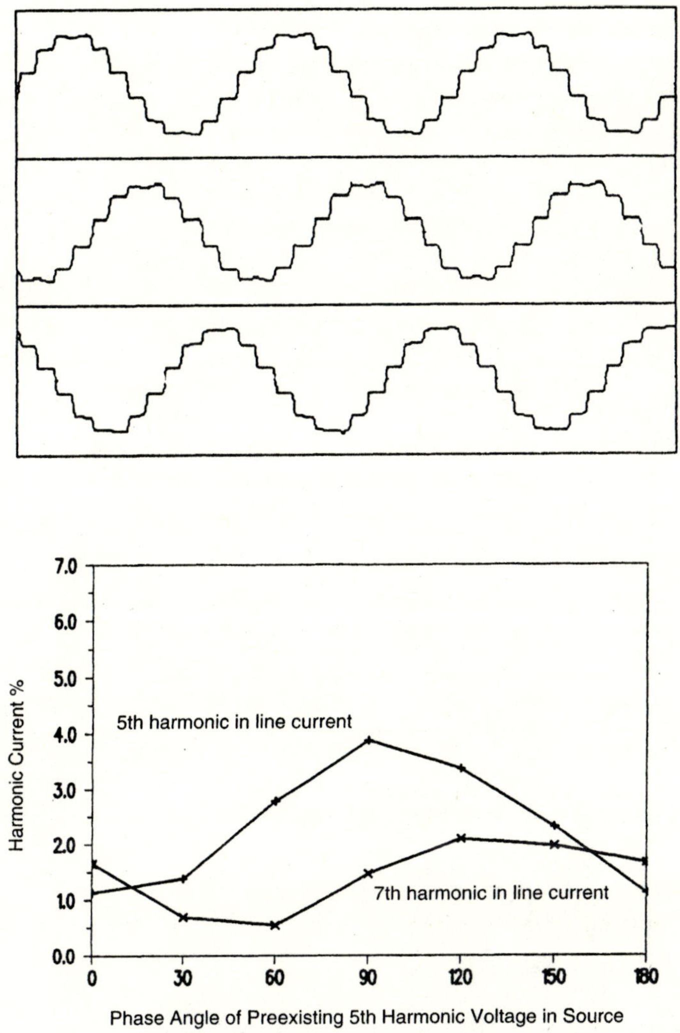

Figure 9-8 Calculated ac line currents and harmonics for 12-pulse circuit, MC-103.

the reduced kVA loading in the wye connection. The main design challenge is to get reasonable balance in the leakage inductances. As previously noted, bifilar windings can be very helpful in this regard.

Turns Ratios. Because the two converters are series connected, the individual dc output and, hence, the required turns ratio of $\sqrt{3}$ are not critical. This contrasts with parallel converters, such as in MC-101.

Discussion of Results. Because dc unbalance is not critical, this multipulse connection is observed to be relatively insensitive to the phase angle of preexisting source harmonics. For 2.5% preexisting 5th harmonic, a worst case residual 5th harmonic current of about 4% occurs. This result is in line with previous recommendations made in Chapter 7. There, it was stated that, without information to the contrary, 20% residuals should be assumed for simple multipulse connections that comprise multiple 6-pulse converters.

The 1% negative-sequence voltage in the source produces an average of 1.3% 3rd harmonic current in the line at full load. This is similar to other circuits.

The harmonic constant under worst case conditions was found to be 122. Thus, this circuit will produce about half the total system voltage distortion of a 6-pulse circuit, which has a typical harmonic constant of 240.

Taking everything into consideration, this circuit produces a robust 12-pulse converter that should perform very well in practical power systems.

Table 9-2 provides typical harmonic currents under the worst case type 2 power operating conditions.

TABLE 9-2.

PERCENTAGE HARMONIC CURRENTS IN CIRCUIT MC-103 WITH TYPE 2 SOURCE POWER AND PREEXISTING 5TH AT 90°								
#3	**#5**	**#7**	**#11**	**#13**	**#17**	**#19**	**#23**	**#25**
1.2	3.9	1.5	7.4	5.2	1.4	1.1	1.6	1.2
harmonic constant = 122 with exact turns ratio								

9.7 EIGHTEEN-PULSE WITH FORK STEP-DOWN TRANSFORMER, CIRCUIT MC-104

The schematic is shown in Figure 9-9. Digital computer results are shown in Figures 9-10, on page 162, and 9-11, on page 164.

This circuit has the capability of meeting the IEEE Std 519-1992 current harmonic limitations without the addition of any other filters. The waveform of line current drawn from the supply has less than 3% total distortion. Its superior performance is not simply because it is an 18-pulse method. Other 18-pulse circuits comprising three 6-pulse bridges give good but not comparable performance. As discussed earlier in this chapter, the 9-pulse midpoint connection reduces commutation voltage such that only a small amount of ac line reactance is needed to filter out higher harmonic frequencies.

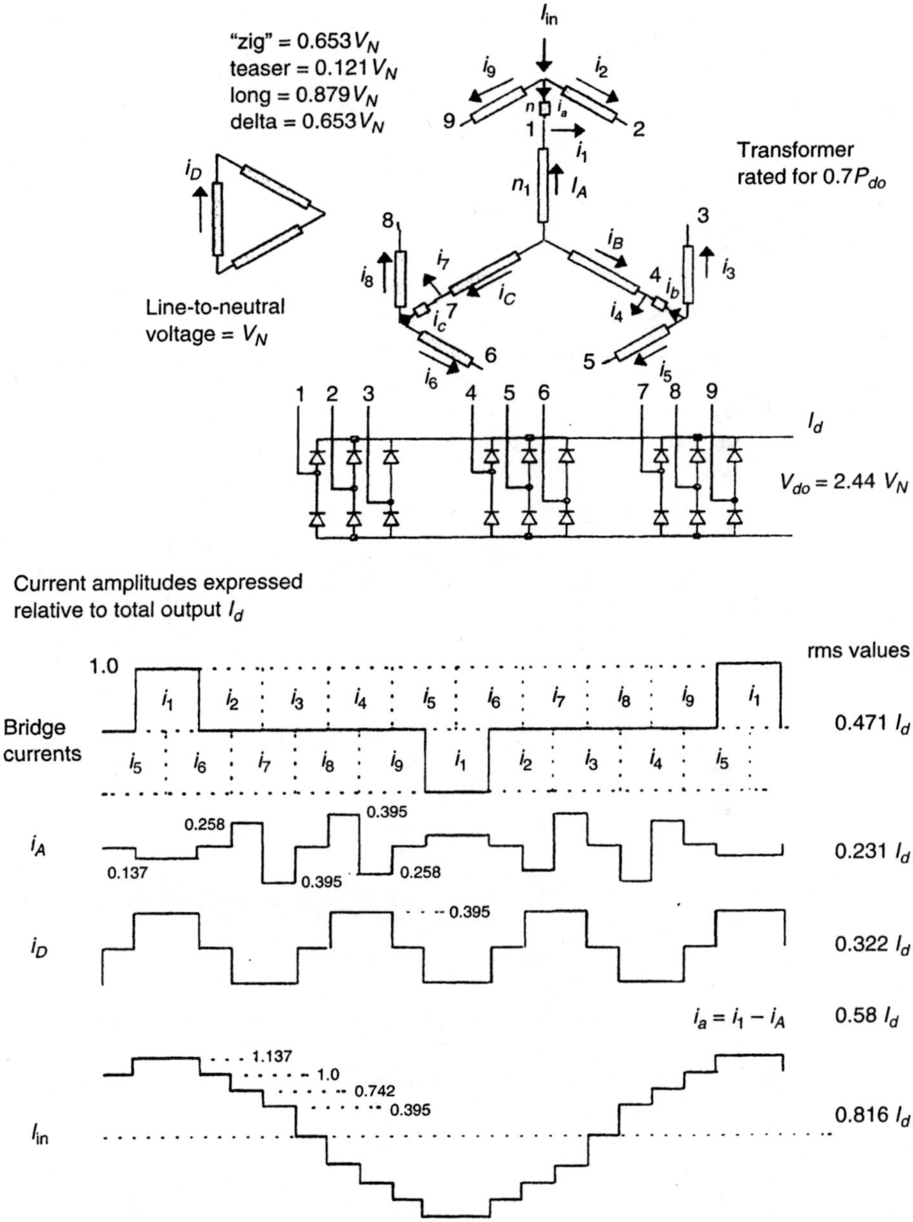

Figure 9-9 Differential fork for an 18-pulse connection, MC-104.

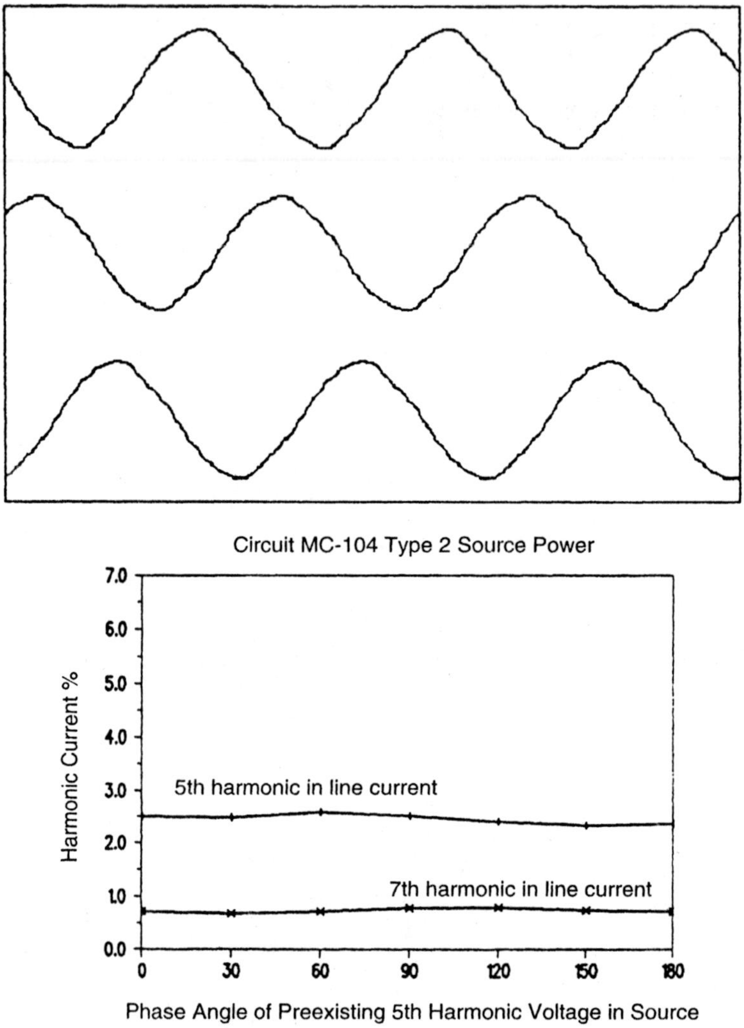

Figure 9-10 Calculated ac line currents and harmonics for an 18-pulse circuit, MC-104.

In the selected design the ac reactance is only 1.44%. Obtaining low reactance is a challenge for the transformer designer. However, when properly configured, the transformer leakages can provide most of the small amount of high-frequency filtering required.

This 18-pulse concept would not be viable without the availability of low-cost high-performance semiconductors. For example, it would have been of little practical value when rectification was obtained using mercury arc converters. This may be one reason that the circuit has not been described elsewhere.

Turns Ratios. Practical turns ratios for this connection are more restricted than those for the comparable differential-delta method shown in Figure 5-8. The ratios listed in Table 9-3 should, however, give good practical performance.

TABLE 9-3.

TURNS RATIOS FOR STEP-DOWN FORK TRANSFORMER FOR 18-PULSE CONVERTER				
Zig	16	22	60	97
Teaser	3	4	11	18
Long	22	29	80	131
Amplitude error	−0.3%	−0.35%	+0.16%	−.012%
Phase error	−0.8°	+0.893°	+0.425°	−0.11°

Discussion of Results. This is one of those circuits where the waveforms clearly show the excellent results. The ac line currents are nearly sinusoidal and meet the IEEE Std 519-1992 current distortion requirements under the most arduous conditions.

These computer results are similar to those obtained practically with the 18-pulse differential delta arrangement described in Chapter 5 and illustrated with practical examples in Chapter 10.

As noted earlier, this circuit has not been proven in practical designs. Also, idealized turns ratios were used to develop the results shown in Figure 9-10. However, experience with similar computer simulations gives a high degree of confidence that the practical performance will be close to that predicted, provided transformer leakages are properly coordinated.

Table 9-4 gives tabulated harmonics existing in the computer simulation. The worst case harmonic constant was calculated to be 34. Using practical turns ratios as shown in the following equations, the harmonic currents and harmonic constant were practically unchanged.

$$\text{Zig winding} = 16 \text{ turns}$$

$$\text{Long winding} = 22 \text{ turns}$$

$$\text{Teaser winding} = 3 \text{ turns}$$

$$\text{Delta winding} = 22 \text{ turns}$$

With proper design of leakage inductances, this circuit has the ability to significantly reduce system total voltage distortion and converter harmonic currents. The system voltage distortion caused by these 18-pulse ac-to-dc power converters will be less than one-sixth that caused by conventional 6-pulse converters.

Figure 9-11 illustrates how this type of 18-pulse circuit continues to give near sinusoidal ac line current at reduced loads. The full-load power factor in

TABLE 9-4.

PERCENTAGE HARMONIC CURRENTS IN CIRCUIT MC-104 WITH TYPE 2 SOURCE POWER AND PREEXISTING 5TH AT 60°								
#3	**#5**	**#7**	**#11**	**#13**	**#17**	**#19**	**#23**	**#25**
1.4	2.6	0.72	0.42	0.38	1.27	1.0	0.24	0.29
harmonic constant = 34								

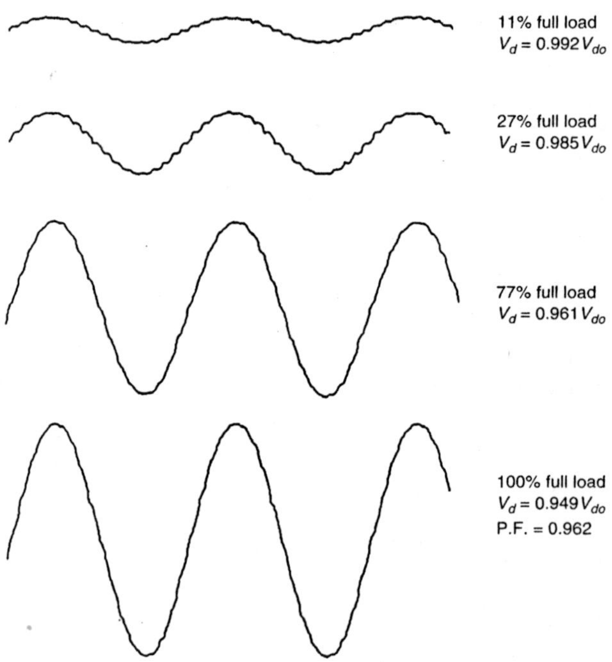

Figure 9-11 Line current and dc voltage for various loads, MC-104 with type 1 power source.

this 112-kW design example is greater than 0.96. The dc voltage drop at full load is less than 5.1%.

This new arrangement and other 18-pulse circuits are expected to find wide application because they enable the limits of IEEE Std 519-1992 to be met.

9.8 TWELVE-PULSE WITH DIFFERENTIAL FORK STEP-DOWN TRANSFORMER, CIRCUIT MC-105

Figure 9-12 shows this new 12-pulse topology. It gives excellent all-round performance and tolerates the effects of type 2 power. With appropriate leakage inductance, the characteristic 11th and 13th harmonics of current can be reduced, a feature difficult to achieve with conventional 12-pulse methods. Shown as an autoconnection in Figure 9-12, it can also be used as a double-wound configuration if power is applied to the delta windings.

Turns Ratios. Practical turns ratios for MC-105 are given in Table 9-5.

TABLE 9-5.

PRACTICAL TURNS FOR THE FORK TRANSFORMER IN MC-105		
N_m **Main**	N_y **Extender**	N_x **Auxiliary**
26	12	7
41	19	11
56	26	15

Computer Calculated Results. Waveforms and harmonic currents with type 1 power are shown in Figure 9-13. Figure 9-14 shows the effect of type 2 power.

Adding Regeneration Features. One method for obtaining regeneration keeps the same polarity load voltage, but reverses the polarity of the load dc current. Such a scheme is shown in Figure 9-15 on page 168. The 12-pulse converter diodes are replaced with thyristors; also, an additional, opposite polarity, 6-pulse thyristor converter is included. Power is returned to the ac source using the opposite polarity converter in conjunction with appropriate gating controls that prevent current from circulating between the converters. In practice, the reliability of the inverting converter is enhanced by raising its ac supply voltage by about 16%. This is easily obtained by adding an extra winding N_r to the fork transformer as shown in Figure 9-15. The extra harmonic

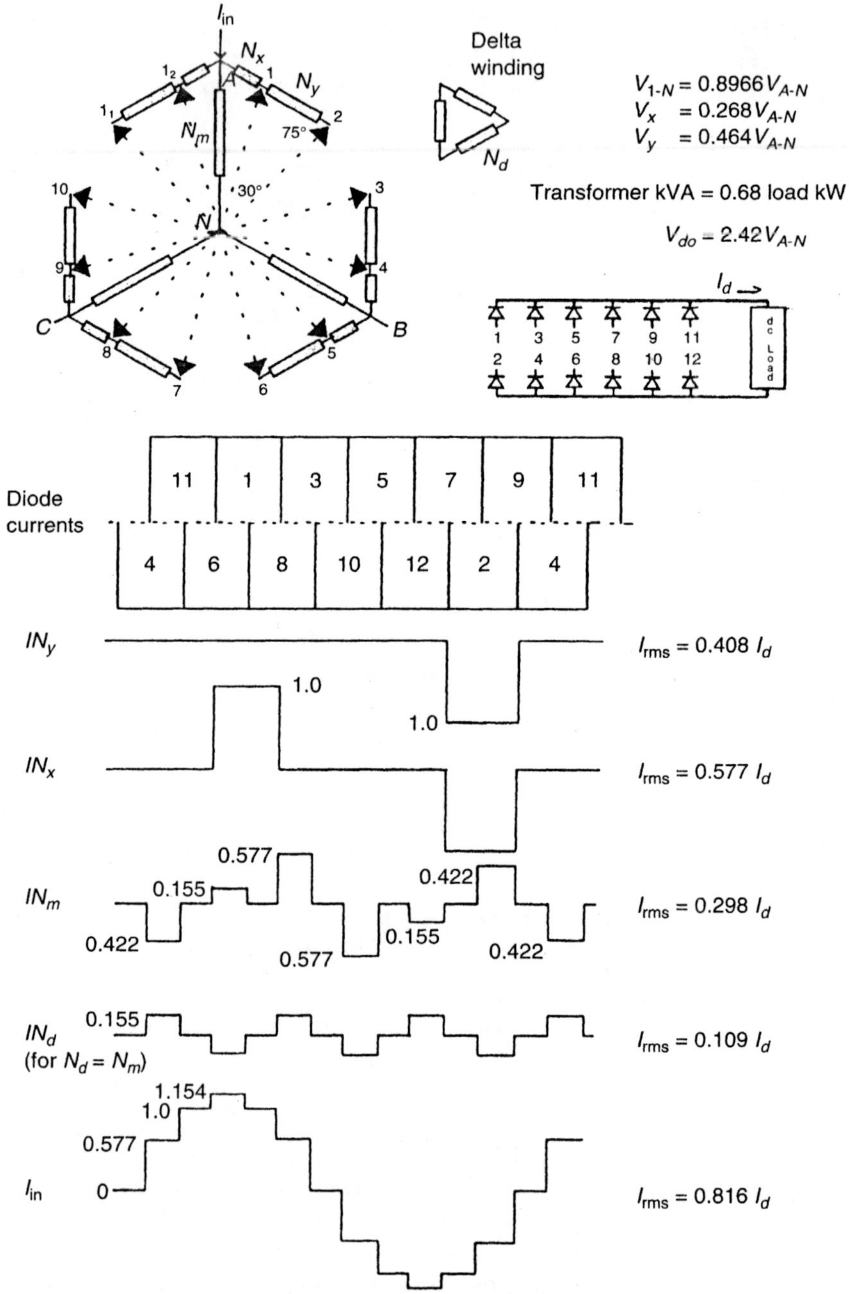

Figure 9-12 Differential fork for 12-pulse connection, MC-105.

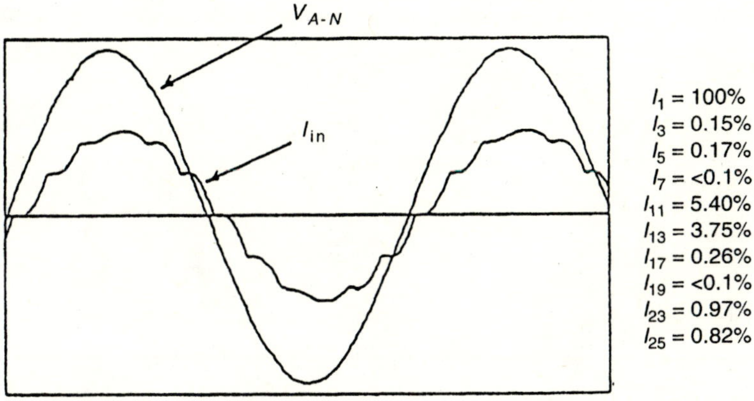

Figure 9-13 Line currents in new 12-pulse connection.

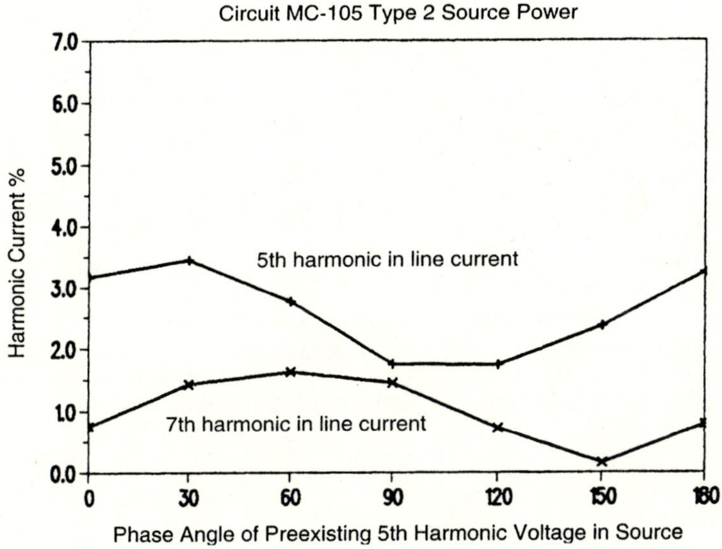

Figure 9-14 New 12-pulse circuit (MC-105) with type 2 power.

currents associated with 6-pulse conversion should not be a concern for loads with intermittent regeneration requirements. Where the regeneration feature is a large part of the duty cycle, the 12-pulse performance can be retained by using a regenerative method which keeps converter current in the same sense, but reverses the polarity of converter voltage [36].

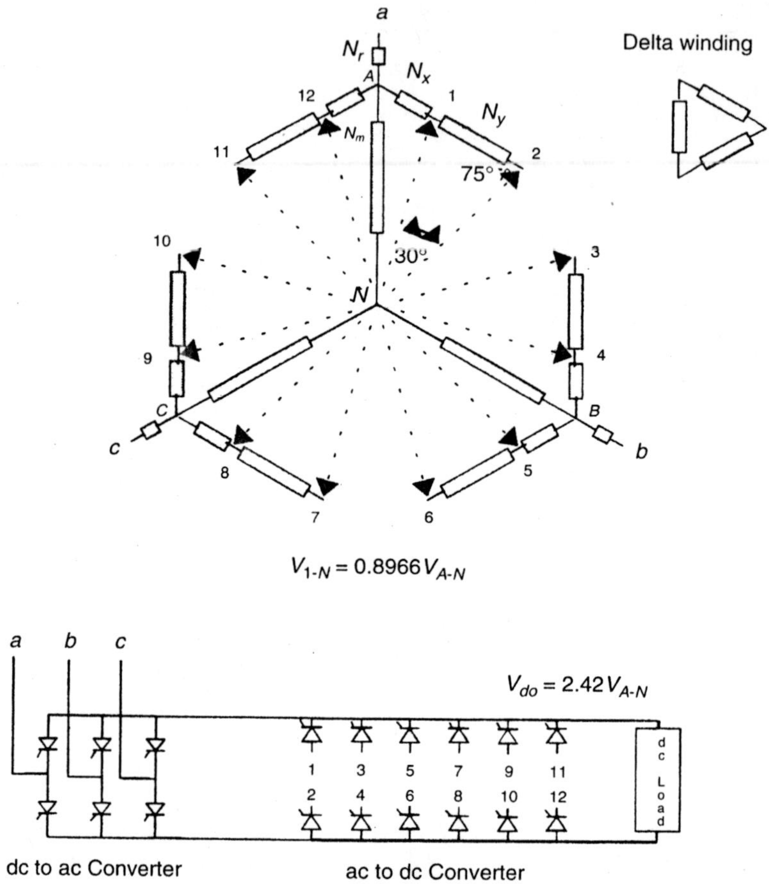

Figure 9-15 Modifications to MC-105 for regeneration.

Chapter 10

Practical Applications

10.1 INTRODUCTION

Any technical work is enhanced by practical results that substantiate the findings, especially when the data is provided by others. This chapter would not have been possible without generous team support. Some of those team players have been noted in the book's acknowledgments.

Converter technical principles are universal, but the ability to present this kind of practical data is unique. I am grateful for the opportunity. Schematics and, where possible, detailed test results are provided. In some cases only the practical waveshape is available, but this is provided to show correlation with the calculations or computer simulation within the text.

10.2 COMPARISON OF CONVERTER TYPES

A very important application of converters is in the "front-end" ac to dc power conversion used in most adjustable-frequency controllers. Typically, a three-phase 480-V, 60-Hz supply provides the power source. Depending upon the system requirements, 6-pulse, 12-pulse, and 18-pulse converters have been used in these equipments. Visual comparison of typical line current waveshapes associated with these methods can be obtained from the oscillograms provided in this chapter.

10.2.1 Six-Pulse with Capacitor-Input Filter

Figure 10-1 displays the line current for a typical 6-pulse converter with capacitor filter. This type of filtering, in conjunction with ac line reactance, has been used up to 150-HP rating but finds greater application in the lower horsepower ratings.

Measured harmonic currents, at full load for one particular installation, are given in Table 10-1. The calculated harmonic constant of 367 is not unusual for this type of equipment when it represents a small fraction of the system load.

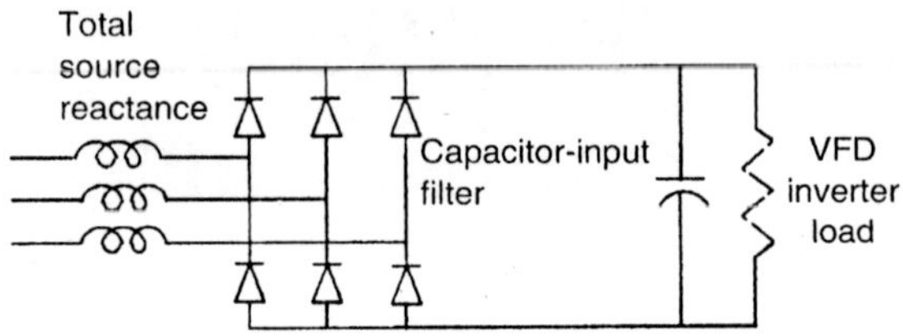

Figure 10-1 Six-pulse converter with capacitor-input filter.

TABLE 10-1.

HARMONIC CURRENTS		
I_3	16.7%	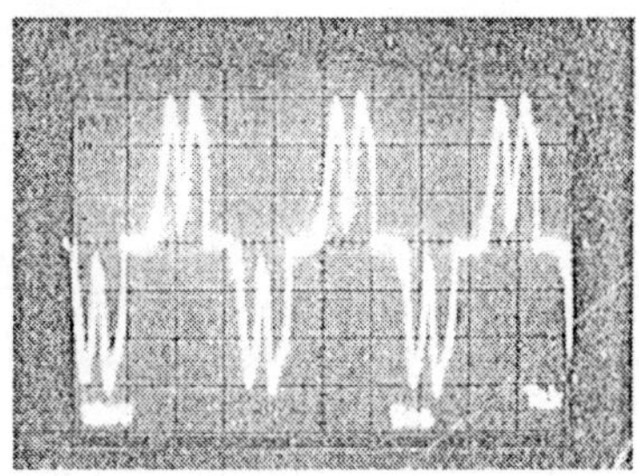
I_5	50.7%	
I_7	29.8%	
I_{11}	8.9%	
I_{13}	6.29%	
I_{17}	3.38%	
I_{19}	1.93%	
I_{23}	2.11%	*THD* (current) = 62.3%
I_{25}	1.51%	percentage harmonic constant = 367

10.2.2 Six-Pulse with Inductor-Input Filter

Figure 10-2 gives a schematic and Table 10-2 provides results for this arrangement. The 5th harmonic of current is seen to be noticeably reduced by the inclusion of a dc inductor. The harmonic constant has been reduced to 256 as well.

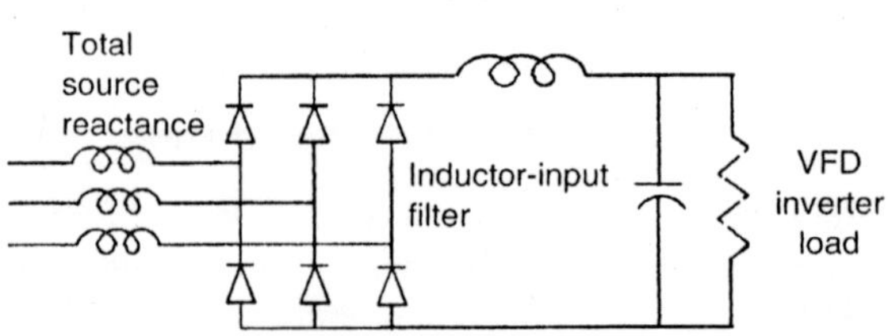

Figure 10-2 Six-pulse converter with inductor-input filter.

TABLE 10-2.

HARMONIC CURRENTS		
I_3	3.78%	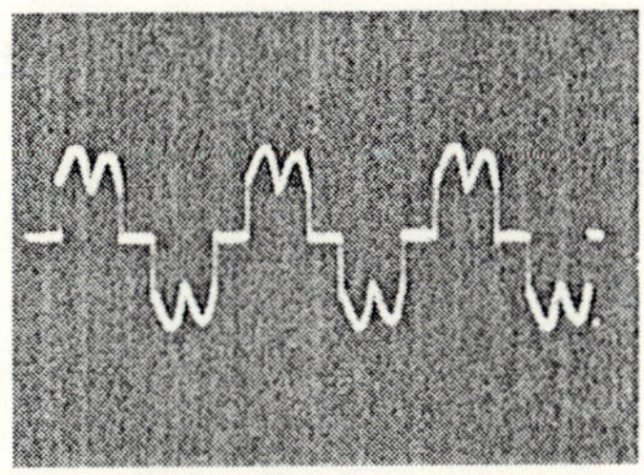
I_5	27.5%	
I_7	10.9%	
I_{11}	9.5%	
I_{13}	3.9%	
I_{17}	5.5%	
I_{19}	3.5%	
I_{23}	3.9%	*THD* (current) = 32.7%

10.2.3 Twelve-Pulse with Autotransformer

This arrangement, shown in Figure 10-3, uses a ±15° polygon-connected autotransformer, two interphase transformers, and 12 SCRs. The dc output voltage is controlled by gating the SCRs appropriately. Typical results near full output are given in Table 10-3 on page 172.

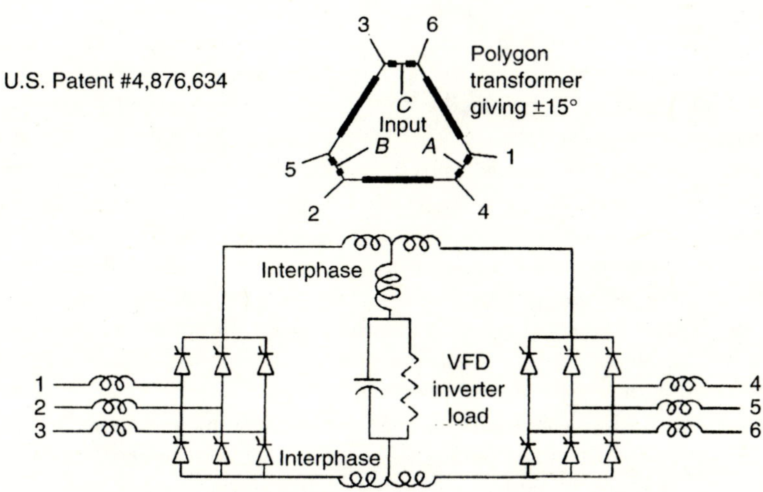

Figure 10-3 Twelve-pulse converter with SCRs.

TABLE 10-3.

HARMONIC CURRENTS		
I_3	1.2%	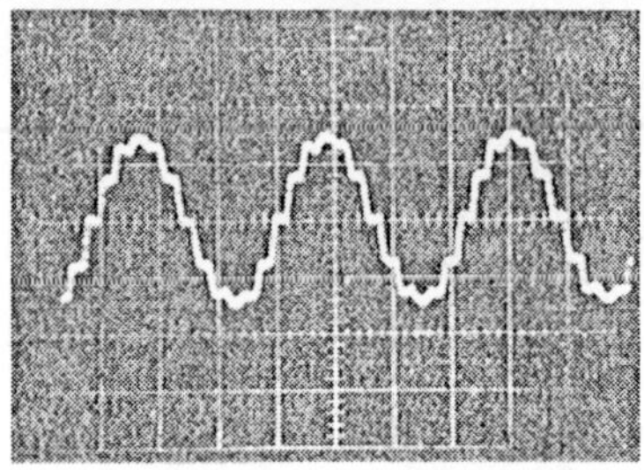
I_5	1.2%	
I_7	0.6%	
I_{11}	7.6%	
I_{13}	1.1%	
I_{17}	0.3%	
I_{19}	0.3%	
I_{23}	1.3%	*THD* (current) = 8.1%
I_{25}	1.1%	percentage harmonic constant = 94.7

10.2.4 Eighteen-Pulse Converter

Figure 10-4 shows the basic schematic of a patented 18-pulse arrangement. This arrangement provides a smooth dc voltage to power a fixed dc voltage, pulse-width-modulated (PWM) variable-frequency drive.

The phase-shifting transformer is similar to that shown in Figure 5-8 except that a small extension is made to the delta winding. This reduces the voltage such that it is the same as obtained with a straight-through, 6-pulse arrangement. Transformer leakage inductance provides high-frequency filtering of the ±40° three-phase voltage sets. A small, physically separate inductor is used to provide equivalent filtering of the high-frequency currents in the 0° set of three-phase voltages.

The line harmonic currents in this connection are greatly reduced compared to 6-pulse and 12-pulse connections. The equipment can meet the IEEE Std 519–1992 harmonic current recommendations under all practical conditions.

Table 10-4 tabulates harmonic currents measured in the first prototype, a 250-kVA unit with 18 SCRs and phase control. Subsequent production units were constructed with diodes and smaller high-frequency filter inductors. Because of its exemplary low-harmonic performance, the design is referred to as "clean power." A photograph of a complete 500-HP VFD is given in Figure 10-5.

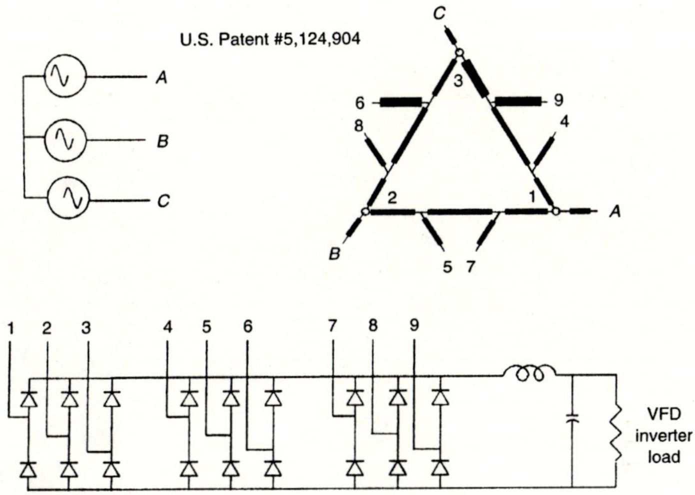

Figure 10-4 Eighteen-pulse "clean power" converter.

TABLE 10-4.

HARMONIC CURRENTS		
I_3	0.7%	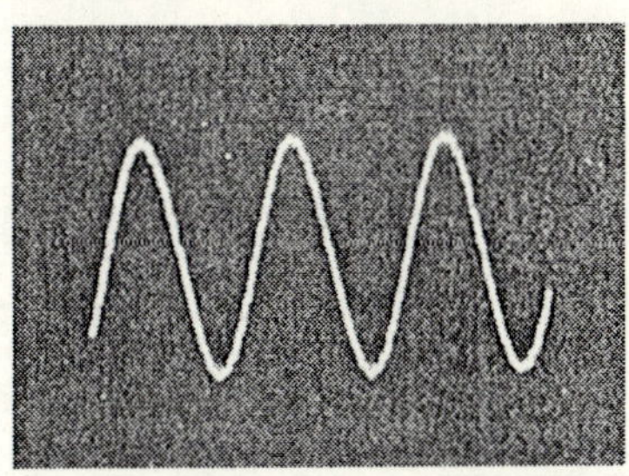
I_5	0.5%	
I_7	0.1%	
I_{11}	0.1%	
I_{13}	0.1%	
I_{17}	0.9%	
I_{19}	1.5%	
I_{23}	0.1%	*THD (current) = 1.97%*
I_{25}	0.1%	percentage harmonic constant = 34

Figure 10-5 500-HP VFD with 18-pulse "clean power" diode converter. Differential-delta $\pm 40^\circ$ transformer with front cover removed. (Courtesy of Westinghouse Electric Corporation, Oldsmar, Florida.)

10.2.5 Tests on Other Topologies

Two candidate 12-pulse circuits are shown in Figures 10-6 and 10-7. The oscillograms display the ac input line current. Figure 10-6 shows the input line current in a half-power delta/wye 12-pulse arrangement. The test unit was rated at 100 HP. As noted in Chapter 2, the half-power transformer rating is feasible only if means are provided to balance the converter currents. Under the test conditions, about 75% of the load was carried by one converter.

Figure 10-7 shows input line current in a 12-pulse method using two auto-connected polygon transformers that provide ($\pm 15^\circ$) phase shift. Good balance was achieved under the given (100-HP) test conditions; however, compensation for unbalance effects is recommended, as discussed in Chapter 2.

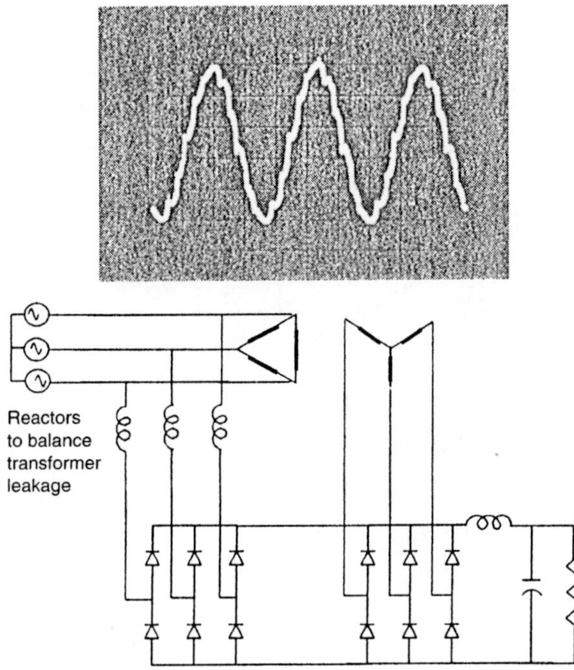

Figure 10-6 Line current in 12-pulse with delta/wye power transformer and without interphase transformer.

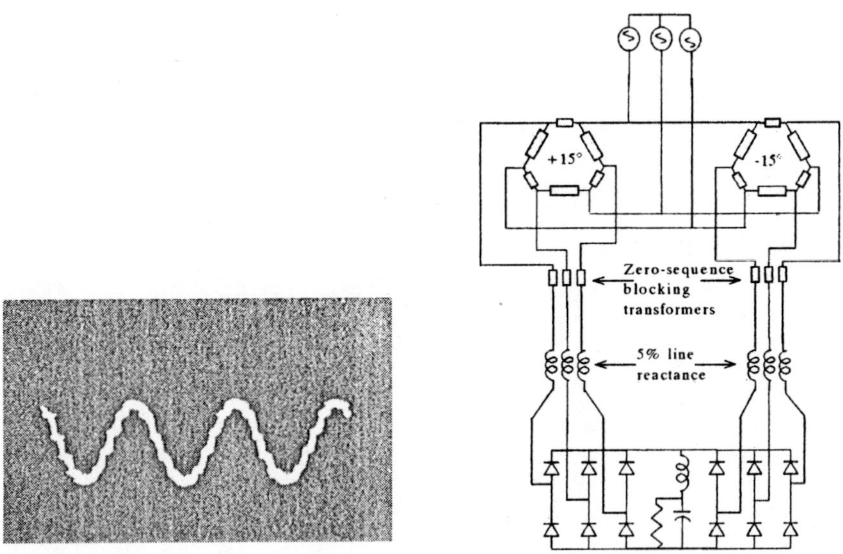

Figure 10-7 Twelve-pulse with two auto-connected polygon phase-shifting transformers, two ZSBTs, and two three-phase ac line reactors.

10.3 FIELD TEST RESULTS

Test results for a 350-HP VFD incorporating an 18-pulse clean power converter are included in Figures 10-9 and 10-10. The equipment is installed and operational.

10.3.1 Analysis of Field Test Results

The equipment under test was a 350-HP variable-frequency drive. The ac-to-dc converter was an 18-pulse clean power type. Construction was generally in line with the unit shown in Figure 10-5. The customer measured and recorded voltage and current harmonics of each phase. The phases were adequately balanced. A summary of the electrical operating conditions is given in Figure 10-8.

The power to the converter had a preexisting total harmonic distortion of approximately 2.7%. Of this distortion, 2.5% was the 5th harmonic. This supports the recommendation to use 2.5% 5th harmonic for type 2 power. Figures 10-9 and 10-10 show the results from phase A. Similar results are obtained from the other phases, although slightly more (5%) 5th harmonic of current was present in phase B.

Under the test conditions, which correspond to about 90% load, the average amount of 5th harmonic current was 4.52%. The largest characteristic harmonic was the 17th, with an amplitude of 2.1%. Under full load conditions the 5th harmonic is estimated as 4.18% and the 17th as 1.89%. To meet the most difficult

```
===========================================
2136 P203 4W              Oct 06 1993 (Wed)

VOLTAGE & CURRENT SNAPSHOT 10:10:05 AM

  VOLTAGE                     279.4 Vrms
  ---------------------------------
    Phase A-N: 279.1 Vrms,      0°(ref)
    Phase B-N: 279.3 Vrms,   -120°
    Phase C-N: 279.9 Vrms,    120°
    Neut-Gnd:    0.0 Vrms,      0°
    Imbalance:   0.2%

  CURRENT                   1.008 kA rms
  ---------------------------------
    Phase A:    343.1 A rms, -13°
    Phase B:    339.1 A rms,-133°
    Phase C:    326.0 A rms,  98°
    Neutral:     48.2 A rms, 174°
    Imbalance:   3.0%
```

Figure 10-8 Summary of field test conditions.

IEEE Std 1992 limits, namely, when $I_{sc}/I_L < 20$, about 28% linear system load is also needed. IEEE Std 519 allows larger characteristic harmonic currents for multipulse inverters, but reduces allowable noncharacteristic harmonic currents, which depend greatly upon the source preexisting voltage harmonics. Many practical systems have significant preexisting harmonic voltages and, for these cases, multipulse inverters can probably best meet harmonic current requirements by simply conforming to the limits in Table 1-2.

2136 P203 4W Oct 06 1993 (Wed)

PHASE A CURRENT SPECTRUM 10:08:08 AM

Fundamental amps: 342.8 A rms

Fundamental freq: 60.0 Hz

HARM	PCT	SINE PHASE	HARM	PCT	SINE PHASE
FUND	100.0%	-13°	2nd	1.2%	138°
3rd	1.0%	33°	4th		
5th	3.9%	-46°	6th		
7th	2.1%	77°	8th		
9th	0.1%	49°	10th		
11th	1.3%	145°	12th		
13th	0.6%	-164°	14th		
15th	0.1%	-45°	16th		
17th	2.1%	-45°	18th		
19th	1.5%	-51°	20th		
21st			22nd		
23rd	0.6%	-79°	24th		
25th	0.6%	-77°	26th		
27th			28th		
29th	0.4%	-4°	30th		
31st	0.2%	17°	32nd		
33rd			34th		
35th	0.4%	177°	36th		
37th	0.2%	-163°	38th		
39th			40th		
41st	0.3%	139°	42nd		
43rd	0.3%	-33°	44th		
45th			46th		
47th	0.1%	-158°	48th		
49th	0.2%	168°	50th		
ODD	5.5%		EVEN	1.2%	

THD: 5.7%

Figure 10-9 Harmonic currents in phase A.

```
2136 P203 4W              Oct 06 1993 (Wed)

PHASE A SNAPSHOT                 10:10:23 AM

Phase A-N VOLTAGE:   279.1 Vrms
                       1.4 Crest Factor
                       1.1 Form Factor

Phase A CURRENT:     343.1 A rms
                       1.5 Crest Factor
                       1.1 Form Factor

CURRENT LAGS VOLTAGE BY 13°(0.97 dPF)
```

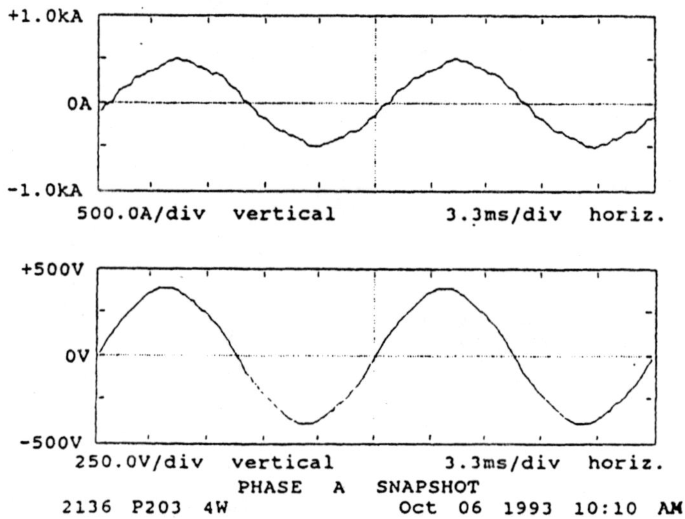

Figure 10-10 Line current and $V_{L\text{-}N}$ phase A.

Chapter 11

Digital Computer Simulations

11.1 INTRODUCTION

The main text has addressed different types of power electronic equipment and presented graphical data that will support design and performance calculations for many of the more straightforward electronics applications. However, several references have been made throughout the text to the need for digital computer simulation in more difficult applications. Means for digital simulation will now be provided for a 6-pulse diode bridge converter fed directly and also through a 1:1 delta/wye transformer. This will give some understanding of the basics of digital simulation techniques and will pave the way for digital analysis of larger, more complex configurations.

11.2 THE SOFTWARE

For digital analysis of power electronic circuits, I have found software called electronic circuit analysis (ECA-2) to be very satisfactory. Good results have been obtained in a reasonable time with a modest 386 33-MHz computer. Purchasing information is given on page xiii of the Preface.

The ECA-2 software is not difficult to use, yet it can be applied to elaborate circuits. For example, a system with one 18-pulse converter and two different 6-pulse converters is modeled with fewer than 140 lines of description. For the book, many calculations were made on version 2.4; however, the software has now (June 1994) advanced to revision 2.65. Detailed operating descriptions are given in the manual supplied with the software, and some student exercises are included. Use of the software to simulate a 6-pulse rectifier bridge is described in this chapter.

11.3 DEFINING THE CIRCUIT

In preparation for analysis, a 6-pulse rectifier circuit is drawn as shown in Figure 11-1. It is not always essential to include practical loss components such as the resistance shown in series with each inductor, however, it is good practice. It may help damp high-frequency oscillations, and it is certainly representative of practical applications.

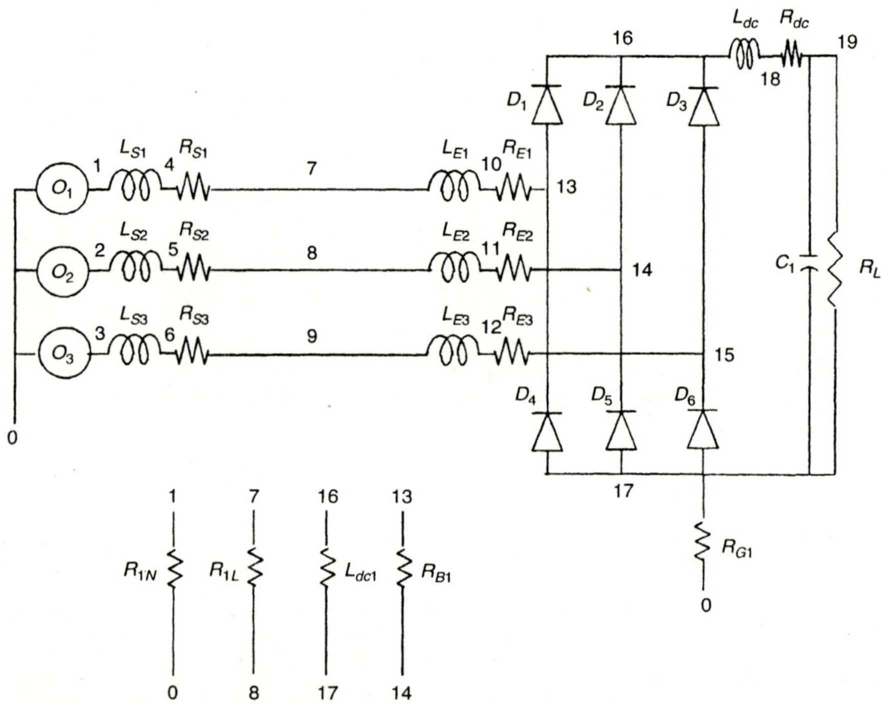

Figure 11-1 Six-pulse bridge rectifier for computer simulation (voltage sensors are defined to probe voltages).

Referring to Figure 11-1, individual components are appropriately labeled, and then node numbers are assigned to each point of connection. The circuit must then be described in a network list. An orderly sequence with multiple comment lines will greatly help in future recognition.

To describe these components in a network list, the program ECA is entered and the command "B" (for build) is given at the prompt. The program keeps advancing line numbers until there are no more components to define. This is determined when a "return" is entered instead of another component. In this

branch	label	nodes		value	function
1	'BOOK1 IS 6-PULSE BRIDGE				
2	'WITH TYPE 1 POWER				
3	'THREE 60-HZ OSCILLATORS				
4	O1	1	0	60.	< 0.
5	+			391.	V 0.
6	O2	2	0	60.	<-120.
7	+			391.	V 0.
8	O3	3	0	60.	< 120.
9	+			391.	V 0.
10	'SOURCE IMPEDANCE				
11	LS1	1	4	68.u	
12	LS2	2	5	68.u	
13	LS3	3	6	68.u	
14	RS1	4	7	0.001	
15	RS2	5	8	0.001	
16	RS3	6	9	0.001	
17	'EQUIPMENT IMPEDANCE				
18	LE1	7	10	68.u	
19	LE2	8	11	68.u	
20	LE3	9	12	68.u	
21	RE1	10	13	0.001	
22	RE2	11	14	0.001	
23	RE3	12	15	0.001	
24	'DIODE BRIDGE				
25	D1	13	16	0.001	F 0.
26	D2	14	16	0.001	F 0.
27	D3	15	16	0.001	F 0.
28	D4	17	13	0.001	F 0.
29	D5	17	14	0.001	F 0.
30	D6	17	15	0.001	F 0.
31	'DC LOAD CIRCUIT				
32	LDC	16	18	800.u	
33	RDC	18	19	0.01	
34	C1	19	17	0.0096	
35	RL	19	17	4.5	
36	RG1	17	0	1.M	
37	'VOLTAGE SENSORS				
38	R1N	1	0	100.K	
39	R1L	7	8	100.K	
40	RDC1	16	17	100.K	
41	RB1	13	14	100.K	

Figure 11-2 Network List for the Circuit in Figure 11-1.

example, we have defined a three-phase source (O1, O2, O3) with individual peak phase voltages of 391 V. This is representative of type 1 power. The resulting network, as listed in the ECA program, is given in Figure 11-2; it has been saved under the name "Book1." It should be noted that typewritten input may be "played back" in a program listing with spaces and decimals given in somewhat different format to those typed in.

11.4 RUNNING THE CIRCUIT

To operate the program, we must first instruct the computer as to where we need to get results. For example, if we type

```
PLOT P2 I LS1 P2 RA -400 400
```

this will set the probe P2 to calculate the current in inductor LS1, and plot it on a graph with scales ranging from −400 A to 400 A.

```
PLOT P1 V R1L P1 RA -800 800
```

will set the probe P1 to plot the voltage across resistor R1L, which is connected across nodes 7 and 8 to monitor the system line-to-line voltage. A maximum of four such probes, P1, P2, P3, and P4, can be in use at any time.

Now we have to decide on the time range of interest and start the operation. It is a 60-Hz power source, so let's assume that we are interested in the final steady-state results. It may take 100 msecs or more for the circuit to settle down, but we would like to watch and see how it progresses to its final values. If we type

```
TRA 0 0.05 50μ Stiff (hit return)
```

the computer interprets this as a transient analysis, starting at time = 0 and going to 0.05 sec in 50-μsec steps. Calculations are made using a "stiff" integrating method, which I have found is less likely to give high-frequency oscillations, or fail to converge, for the type of circuits analyzed in this book. The screen will display a graph for the selected plots and will run until the time has elapsed. When the prompt has returned, operation can proceed for the next 0.05 sec with the same time steps by simply typing

```
TRA STIFF (hit return)
```

If you want to change component values, or time steps, or stopping time, or what you are plotting, you can do this at any prompt.

If you want to modify the circuit by adding components, you can go ahead and type "B." This will give more line numbers and the new components can be added; however, note that because it is a new circuit, the next analysis will restart at time zero.

Results from a "run" on the circuit after 0.3 sec has elapsed, and the start-up transient has decayed, are given in Figure 11-3.

These results show some expected waveforms and help provide understanding of the circuit operation.

To obtain printed copy of the calculated displays a screen-capture program is easy to use. The waveforms shown here were obtained using "GRAFPLUS" software, available from Jewell Industries in Aylesbury, England.

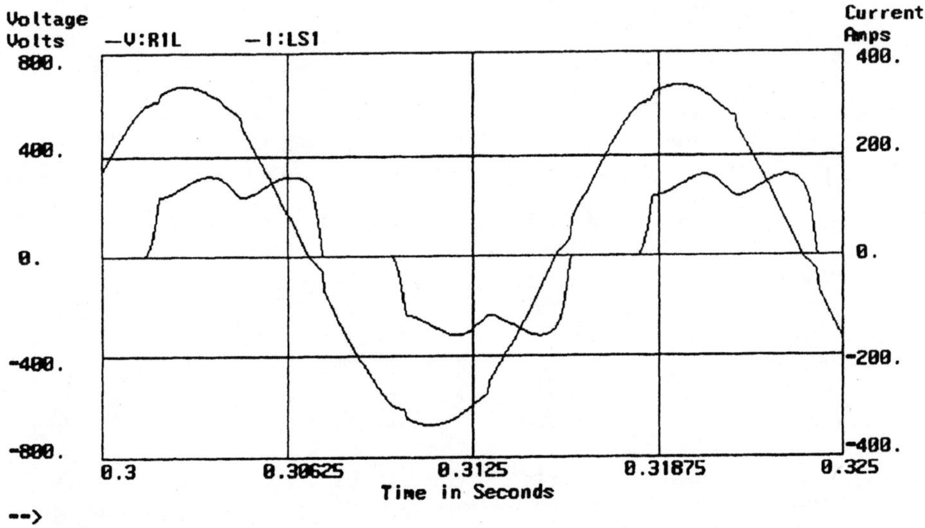

Figure 11-3 Results from a "run" on the circuit in Figure 11-1, after the start-up transient has been damped.

11.5 FOURIER ANALYSIS

To get numerical results for the harmonic distortion of the line current in inductor LS1, it is necessary to perform a Fourier analysis. For this it is required to set up probes at the point of interest. For example,

```
PROBE I LS1
```

will set up to probe the current in line inductance LS1.

The instruction

```
FOURIER 60 1500 60 SKIP 10 STIFF
```

will cause the Fourier routine to proceed. It starts at 60 Hz and proceeds to 1500 Hz in 60-Hz increments. Samples for the fast Fourier transform calculation are determined at intervals of 260.4 μsec. Time steps for the transient calculation are at one-tenth of this value, namely, 26 μsec. Results from this analysis are shown in Figure 11-4 for the probed current.

```
  FOURIER 60 1500 60 SKIP 10 STIFF
Fourier analysis using: Transient  0.3           0.316667      260.417u
                        --------- actual ---------   -------- relative --------
freq        probe      value          dB      phase  value          dB     phase
60.         I:LS1      156.31       43.88    -9.068  1.            0.00    0.000
120.        I:LS1      0.1444      -16.81   -37.128  923.77u     -60.69  -28.059
180.        I:LS1      0.14099     -17.02    -9.822  901.97u     -60.90   -0.753
240.        I:LS1      0.038457    -28.30   158.724  246.03u     -72.18  167.793
300.        I:LS1      38.053       31.61   125.362  0.24344     -12.27  134.430
360.        I:LS1      0.048523    -26.28   109.865  310.42u     -70.16  118.933
420.        I:LS1      16.057       24.11   153.560  0.10272     -19.77  162.628
480.        I:LS1      0.01915     -34.36    62.959  122.51u     -78.24   72.027
540.        I:LS1      0.23437     -12.60   151.088  0.0014993   -56.48  160.157
600.        I:LS1      0.020555    -33.74   -92.709  131.5u      -77.62  -83.641
660.        I:LS1      11.943       21.54   -89.120  0.076402    -22.34  -80.052
720.        I:LS1      0.024562    -32.19  -139.378  157.13u     -76.07 -130.310
780.        I:LS1      7.7557       17.79   -89.205  0.049617    -26.09  -80.136
840.        I:LS1      0.0091485   -40.77   163.310  58.527u     -84.65  172.378
900.        I:LS1      0.33089      -9.61   -82.774  0.0021168   -53.49  -73.706
960.        I:LS1      0.015219    -36.35    18.367  97.362u     -80.23   27.435
1.02K       I:LS1      5.7439       15.18    41.100  0.036747    -28.70   50.169
1.08K       I:LS1      0.01725     -35.26   -29.148  110.36u     -79.14  -20.080
1.14K       I:LS1      4.4322       12.93    38.755  0.028355    -30.95   47.824
1.2K        I:LS1      0.0059151   -44.56   -97.958  37.842u     -88.44  -88.890
1.26K       I:LS1      0.4508       -6.92    27.247  0.002884    -50.80   36.315
1.32K       I:LS1      0.01296     -37.75   128.810  82.914u     -81.63  137.878
1.38K       I:LS1      2.7368        8.74   165.776  0.017509    -35.13  174.844
1.44K       I:LS1      0.014068    -37.04    81.086  90.001u     -80.92   90.154
1.5K        I:LS1      2.5389        8.09   166.815  0.016243    -35.79  175.883
-->
```

Figure 11-4 Fourier results for Figure 11-1.

11.6 INCORPORATING TYPE 2 POWER SOURCE

To incorporate type 2 power source with 1% negative sequence voltage, and 2.5% of 5th harmonic voltage, the individual oscillator voltages must be modified. This has been done in the abbreviated listing shown in Figure 11-5. Note that these parameters are simply added as a "+" extension line, with the correct frequency, phase position, and amplitude, as required. In the particular case shown, the negative sequence and preexisting 5th harmonic voltages are in phase with the fundamental voltage. As discussed in Section 2.6.2, it may be necessary to evaluate variations of these phase relationships in some analyses.

branch	label	nodes		value	function
4	01	1	0	60.	< 0.
5	+			391.	V 0.
6	+			60.	< 0.
7	+			3.91	V 0.
8	+			300.	< 0.
9	+			9.78	V 0.
10	02	2	0	60.	<-120.
11	+			391.	V 0.
12	+			60.	< 120.
13	+			3.91	V 0.
14	+			300.	< 120.
15	+			9.78	V 0.
16	03	3	0	60.	< 120.
17	+			391.	V 0.
18	+			60.	<-120.
19	+			3.91	V 0.
20	+			300.	<-120.
21	+			9.78	V 0.

Figure 11-5 Network list changes to include type 2 power.

11.7 INCORPORATING A DELTA/WYE TRANSFORMER

There are different ways in which transformers can be simulated. For simple arrangements such as a 1:1 delta/wye connection, an ideal transformer can be represented as described in the ECA-2 manual. Using this technique a modified 6-pulse rectifier connection can be modeled as shown in Figure 11-6. Sample results from this circuit are shown in Figures 11-7 and 11-8. Figure 11-9 on page 188 provides the network list for the schematic shown in Figure 11-6. A separate file saved as TX2.MOD has been set up to define the ratio of a single-phase transformer TX2. Three of these transformers combine to give a 1:1 delta/wye connection.

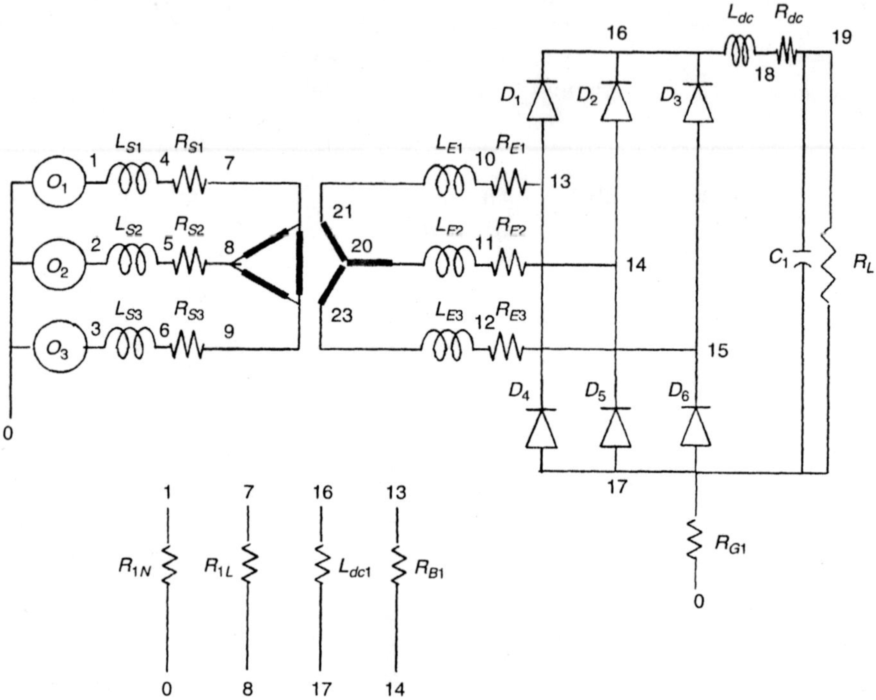

Figure 11-6 Six-pulse rectifier with 1:1 delta/wye transformer.

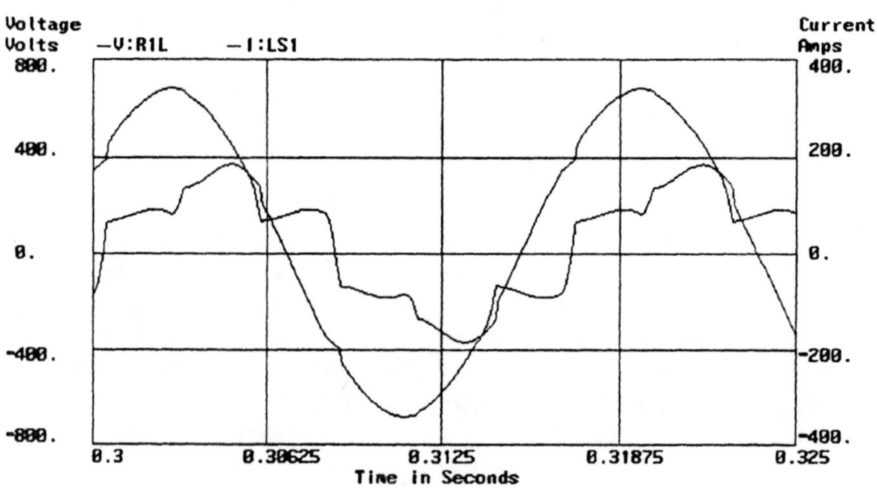

Figure 11-7 Sample result from a "run" on the circuit in Figure 11-6.

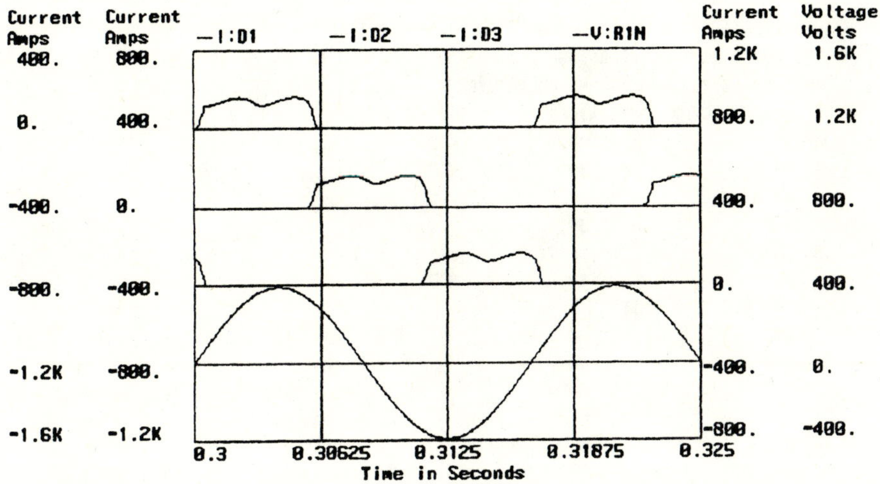

Figure 11-8 Sample result from "runs" on the circuit in Figure 11-6.

11.8 SUMMARY

The foregoing brief discussion is no more than a basic introduction, but it should suffice to illustrate the simplicity and power of digital circuit simulations. It is as if the user has access to a fully equipped power electronics laboratory with all sorts of sophisticated measuring instruments that can be probed anywhere in the circuit. Design changes can be made at will, and high-power performance can be measured without risk.

branch	label	nodes		value	function
1	'BOOK2 IS 6-PULSE BRIDGE FED				
2	'WITH TYPE 1 POWER AND DELTA/WYE				
3	'THREE 60-HZ OSCILLATORS				
4	01	1	0	60.	< 0.
5	+			391.	V 0.
6	02	2	0	60.	<-120.
7	+			391.	V 0.
8	03	3	0	60.	< 120.
9	+			391.	V 0.
10	'SOURCE IMPEDANCE				
11	LS1	1	4	68.u	
12	LS2	2	5	68.u	
13	LS3	3	6	68.u	
14	RS1	4	7	0.001	
15	RS2	5	8	0.001	
16	RS3	6	9	0.001	
17	'THREE-PHASE 1:1 DELTA WYE				
18	T1	7	8	TX2	
19	*	21	20	0.	
20	T2	8	9	TX2	
21	*	22	20	0.	
22	T3	9	7	TX2	
23	*	23	20	0.	
24	RGT	20	0	10.M	
25	'EQUIPMENT IMPEDANCE				
26	LE1	21	10	68.u	
27	LE2	22	11	68.u	
28	LE3	23	12	68.u	
29	RE1	10	13	0.001	
30	RE2	11	14	0.001	
31	RE3	12	15	0.001	
32	'DIODE BRIDGE				
33	D1	13	16	0.001	F 0.
34	D2	14	16	0.001	F 0.
35	D3	15	16	0.001	F 0.
36	D4	17	13	0.001	F 0.
37	D5	17	14	0.001	F 0.
38	D6	17	15	0.001	F 0.
39	'DC LOAD CIRCUIT				
40	LDC	16	18	800.u	
41	RDC	18	19	0.01	
42	C1	19	17	0.0096	
43	RL	19	17	4.5	
44	RG1	17	0	1.M	
45	'VOLTAGE SENSORS				
46	R1N	1	0	100.K	
47	R1L	7	8	100.K	
48	RDC1	16	17	100.K	
49	RT1	21	20	100.K	
50	RT2	22	20	100.K	
51	RT3	23	20	100.K	
52	RB1	13	14	100.K	

Figure 11-9 Network list for the circuit in Figure 11-6.

Appendix

Useful Formulas for Analysis

TRIANGLES

Right-Angled Triangle

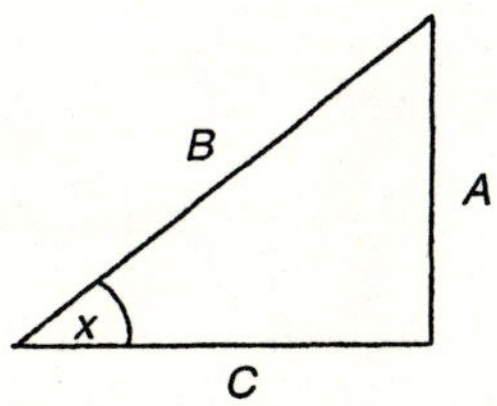

$$\sin x = \frac{\text{opposite}}{\text{hypotenuse}}$$

$$\cos x = \frac{\text{adjacent}}{\text{hypotenuse}}$$

$$\tan x = \frac{\text{opposite}}{\text{adjacent}}$$

General Triangle

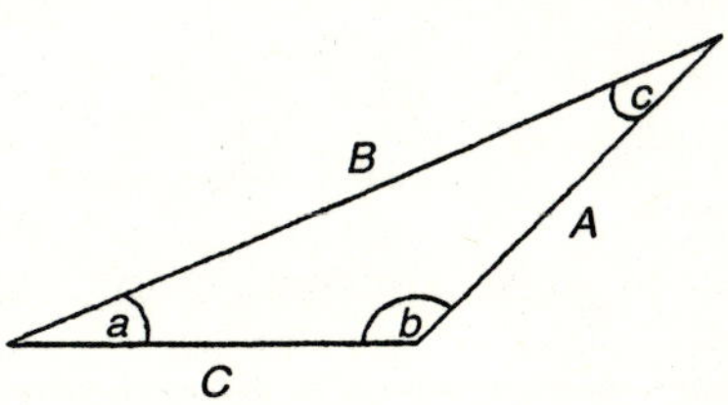

$$\frac{A}{\sin a} = \frac{B}{\sin b} = \frac{C}{\sin c}$$

$$\angle a + \angle b + \angle c = 180°$$

$$\sin(x \pm y) = \sin x \cos y \pm \cos x \sin y$$

$$\cos(x \pm y) = \cos x \cos y \mp \sin x \sin y$$

$$A^2 = B^2 + C^2 - 2BC\cos a$$

$$\sin x + \sin y = 2\sin\frac{1}{2}(x+y)\,\cos\frac{1}{2}(x-y)$$

$$\sin x - \sin y = 2\cos\frac{1}{2}(x+y)\,\sin\frac{1}{2}(x-y)$$

General Triangle, continued

$$\cos x + \cos y = 2\cos\frac{1}{2}(x+y)\cos\frac{1}{2}(x-y)$$

$$\cos x - \cos y = -2\sin\frac{1}{2}(x+y)\sin\frac{1}{2}(x-y)$$

VECTOR OPERATORS

Multiplying a vector by $R\angle\Phi$ multiplies the amplitude by R and rotates the vector an angle Φ in the positive (ccw) direction. Vector operators may be presented in different ways. For example,

$$R\angle\Phi = R(\cos\Phi + j\sin\Phi) = Re^{j\Phi} = a + jb$$

Thus,

$$j = 1\angle 90°; \quad j^2 = 1\angle 180° = -1; \quad j^3 = 1\angle 270° = -j;$$
$$-0.5 + j(\sqrt{3}/2) = 1\angle 120°; \quad -0.5 - j(\sqrt{3}/2) = 1\angle 240°$$

FOURIER SERIES

Some examples of Fourier series are presented that define idealized waveforms of currents and voltages found in various converter circuits. In general,

$$f(\omega t) = \frac{a_0}{2} + a_1\cos\omega t + a_2\cos 2\omega t + a_3\cos 3\omega t + \cdots + b_1\sin\omega t + b_2\sin 2\omega t + \ldots$$

When $f(\omega t)$ is a periodic function with period 2π, then

$$a_n = \frac{1}{\pi}\int_{-\pi}^{+\pi} f(\omega t)\,.\cos n\omega t\, d\omega t; \qquad n = 0, 1, 2, \ldots$$

$$b_n = \frac{1}{\pi}\int_{-\pi}^{\pi} f(\omega t)\,.\sin n\omega t\, d\omega t; \qquad n = 1, 2, \ldots$$

If $f(-\omega t) = f(\omega t)$, the function is even and $b_n = 0$.
If $f(-\omega t) = -f(\omega t)$, the function is odd and $a_n = 0$.
If $f(\omega t + \pi) = -f(\omega t)$, the series contains only odd harmonics.

Example 1

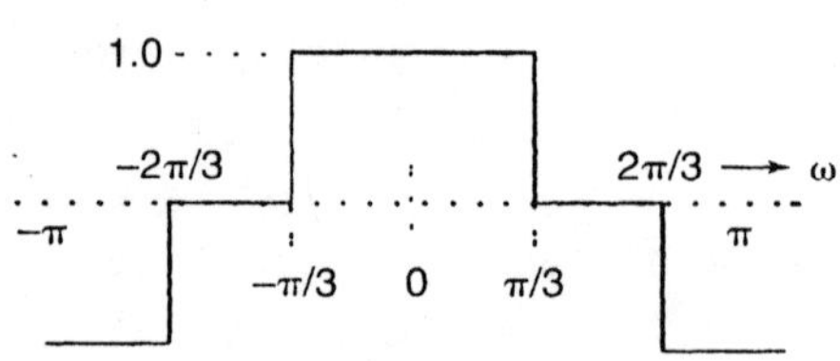

Figure A-1 Idealized 6-pulse waveshape defined about center of unity pulse amplitude.

From inspection of the waveshape shown in Figure A-1, the function has zero average value over the range of $-\pi$ to $+\pi$; thus, $a_0 = 0$. Since $f(-\omega t) = f(\omega t)$, the function is an even function and $b_n = 0$. Thus, only a_n has to be determined. Since cos ωt is an even function, the product $f(\omega t)$ cos ωt is an even function. Therefore, from symmetry we can state

$$a_n = \frac{2}{\pi}\int_{-\pi}^{\pi} f(\omega t)\cos n\omega t \, d\omega t$$

Thus,

$$\frac{\pi a_n}{2} = \int_0^{\pi/3} 1\cos n\omega t \, d\omega t + \int_{\frac{2\pi}{3}}^{\pi} -1\cos n\omega t \, d\omega t$$

This yields

$$a_n = \frac{4}{\pi n}\sin n\frac{\pi}{2}\cos n\frac{\pi}{6}$$

Thus,

$$f(\omega t) = \frac{2\sqrt{3}}{\pi}\left(\cos\omega t - \frac{\cos 5\omega t}{5} + \frac{\cos 7\omega t}{7} - \frac{\cos 11\omega t}{11}\cdots\right)$$

Example 2

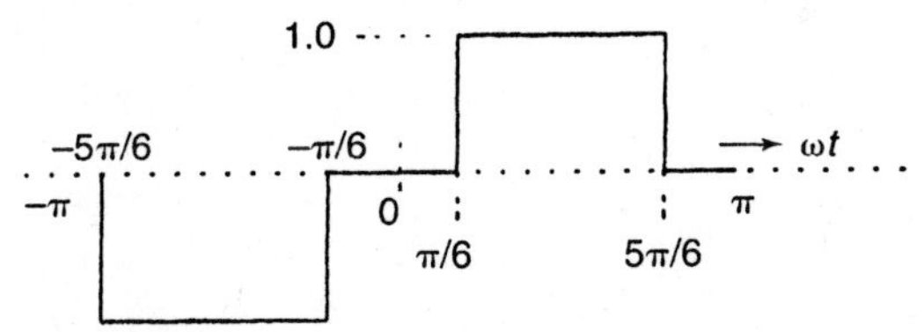

Figure A-2 Idealized 6-pulse waveshape defined about center of zero pulse amplitude.

The waveform depicted in Figure A-2 is of a similar waveform to Figure A-1 but is displaced by an angle of $\pi/2$; thus,

$$f(\omega t) = \frac{2\sqrt{3}}{\pi}\left[\cos(\omega t - \pi/2) - \frac{\cos 5(\omega t - \pi/2)}{5} + \frac{\cos 7(\omega t - \pi/2)}{7} - \cdots\right]$$

$$= \frac{2\sqrt{3}}{\pi}\left[\sin\omega t - \frac{\sin 5\omega t}{5} - \frac{\sin 7\omega t}{7} + \frac{\sin 11\omega t}{11}\cdots\right]$$

Example 3

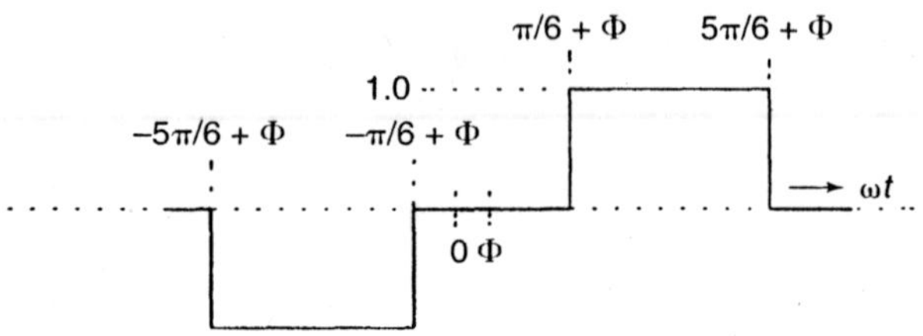

Figure A-3 Idealized 6-pulse waveshape defined at angle Φ.

Figure A-3 shows the same shape of wave as in Figure A-2. In this case, however, the pulse of duration $2\pi/3$ is displaced by an angle Φ to give a more general result.

$$f(\omega t)=\frac{2\sqrt{3}}{\pi}\left[\sin(\omega t-\Phi)-\frac{\sin 5(\omega t-\Phi)}{5}-\frac{\sin 7(\omega t-\Phi)}{7}+\frac{\sin 11(\omega t-\Phi)}{11}\cdots\right]$$

Example 4

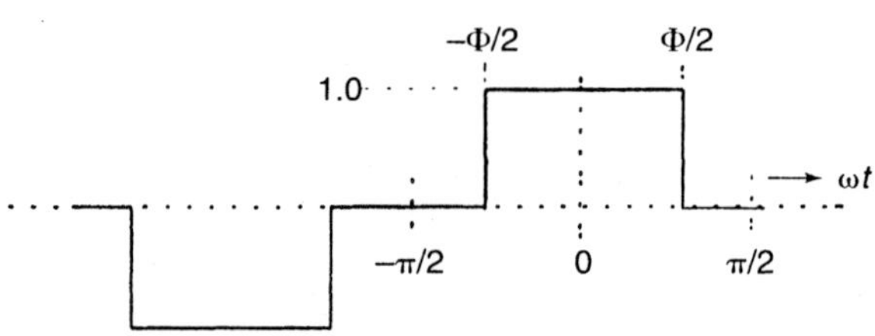

Figure A-4 Idealized pulse of variable width Φ.

In Figure A-4 the width of the pulse is variable. It can range from zero (no output) to $2\pi/3$ for the classical bridge converter current to π for a square wave. Applying the same Fourier principles, the general result is obtained as

$$f(\omega t)=\sum\frac{4}{\pi n}\sin n\frac{\Phi}{2}(\cos n\omega t)$$

where $n = 1, 3, 5, 7, \ldots$

Example 5

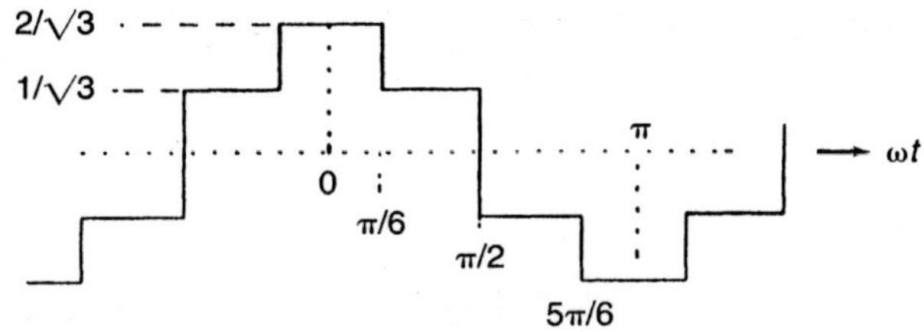

Figure A-5 Idealized 6-pulse wave at input of 1:1 delta/wye transformer.

The wave in Figure A-5 is representative of the line current wave into a 1:1 delta/wye transformer feeding a 6-pulse rectifier load with unity dc current. The fundamental current is identical to that of a rectifier fed without a phase-shifting transformer. The different shape is caused by the different phase angle of the harmonic currents of order $6(2k-1) \pm 1$.

Applying the Fourier techniques, we get

$$a_n = \frac{2}{\pi}\int_0^{\pi} f(\omega t)\ \cos\ n\omega t\ d\omega t$$

from which

$$a_n = \frac{8}{\sqrt{3}\pi n}\left(\sin n\frac{p}{3}\ \cos\ n\frac{p}{6}\right)$$

Thus,

$$f(\omega t) = \frac{2\sqrt{3}}{\pi}\left[\cos\omega t + \frac{\cos 5\omega t}{5} - \frac{\cos 7\omega t}{7} - \frac{\cos 11\omega t}{11}\cdots\right]$$

EXAMPLES OF LINE CURRENT WAVEFORMS

The amplitudes shown in Figure A-6 are given relative to the converter dc current in connections with identical dc output voltage. Similar waveshapes occur in other connections, but "step" amplitudes relative to the dc output current may vary depending upon the inherent dc output voltage.

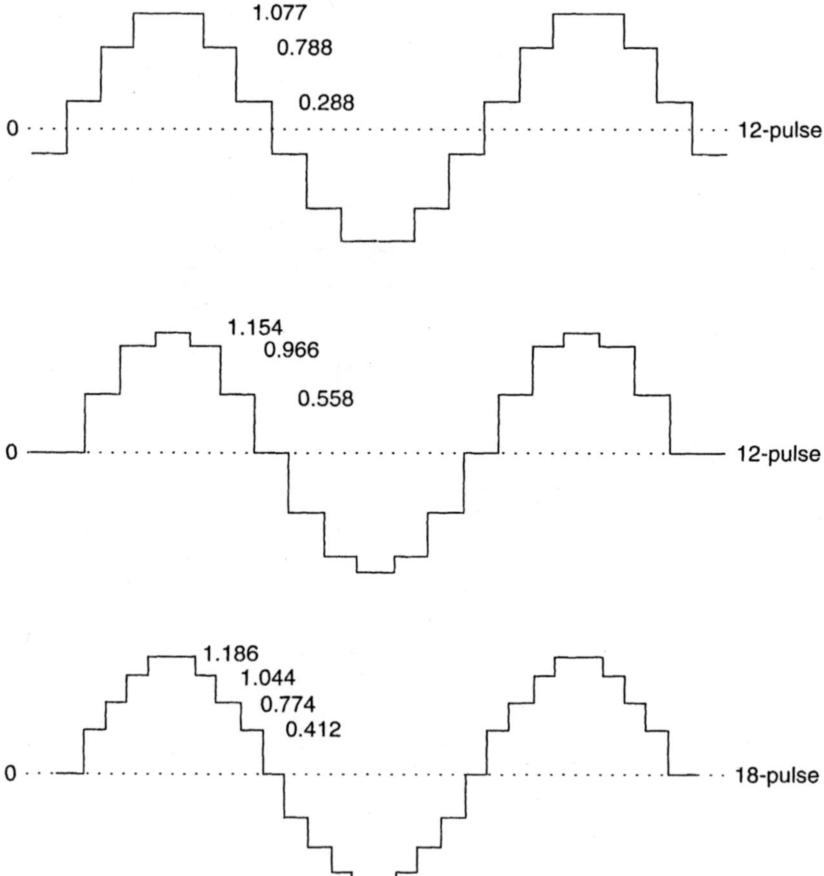

Figure A-6 Examples of typical 12-pulse and 18-pulse waveshapes.

Bibliography

The following selection of reference materials, provided to augment the text discussions, focuses on works that the author himself has found particularly helpful. It is by no means an exhaustive list; however, within the referenced books and papers will be found further references to a large amount of supporting material.

1. Schaeffer, J. *Rectifier Circuits: Theory and Design.* New York: Wiley-Interscience, 1965.
2. Pelly, B. R. *Thyristor Phase-controlled Converters and Cycloconverters.* New York: Wiley-Interscience, 1971.
3. Arillaga, J., D. A. Bradley, and P. S. Bodger. *Power System Harmonics.* New York: Wiley-Interscience, 1985.
4. Bedford, B. D., and R. G. Hoft. *Principles of Inverter Circuits.* New York: John Wiley, 1964.
5. IEEE Std 519-1992. *IEEE Recommended Practices and Requirements for Harmonic Control in Electrical Power Systems.*
6. Depenbrock, M., and C. Niermann. *A New 12-Pulse Rectifier Circuit with Line Side Interphase Transformer.* Lehrstuhl fur Erzeugung und Anwendung Elektrischer Energie, Ruhr Universitat, Postfach 102148, 4630 Bochum 1, West Germany, 1989.
7. Dobinson, L. G. "Closer accord on harmonics," *Electronics and Power,* May 1975, pp. 567–572.
8. Sukegawa, T., et al. "A multiple PWM GTO line side converter for unity power factor and reduced harmonics," *IEEE/IAS Annual Meeting Conference Record,* Dearborn, Michigan, 1991, pp. 279–284.
9. Rosa, J. "Outboard commutation inductors for hexagon converters," U.S. Patent #4,683,257, July 1987.

10. Ludbrook, A. "Harmonic filters for notch reduction." *IAS Transactions,* Vol. 24, no. 5, September/October 1988, pp. 947–954.
11. Moran, S. "A line voltage regulator/conditioner for harmonic sensitive load isolation." *1989 Conference Record of IEEE,* IAS Annual Meeting, San Diego.
12. ANSI/IEEE C57.110-1986. *IEEE Recommended Practice for Establishing Transformer Capability When Supplying Nonsinusoidal Load Currents.*
13. Keskar, P. Y. "PLC system with emergency generators and variable frequency drives: A hard knocks education." *Programmable Controls,* July/August 1988.
14. Corbyn, D. B. "This business of harmonics." *Electronics and Power,* May 1975, pp. 567–577.
15. Hammond, P. W. "A harmonic filter installation to reduce voltage distortion from static power converters." *IEEE Transactions on Industry Applications,* February 1988.
16. Gyugyi, L., and R. A. Otto. "Static VAR Compensation of Electric Arc Furnaces." Paper presented at the VIE Meeting, Liege, Belgium, 1976.
17. Niermann, C. "New rectifier circuits with low mains pollution and additional low cost inverter for energy recovery." Volume III of the *Proceedings of the 3rd European Conference on Power Electronics and Applications,* EPE, Aachen, 1989.
18. Jarc, D. A., and R. G. Schieman. "Power line considerations for variable frequency drives." *IEEE Transactions on Industry Applications,* October 1985.
19. Paice, D. A., and R. J. Spreadbury. "Calculating and controlling harmonics caused by power converters." *1989 Conference Record of IEEE,* IAS Annual Meeting, San Diego, 1989.
20. Central Station Engineers of the Westinghouse Electric Corporation, East Pittsburgh, Pennsylvania. *Electrical Transmission and Distribution Reference Book,* Fourth Edition, 1964, p. 653.
21. Kemp, P. *Alternating Current Engineering,* pp. 633–636. London: Macmillan, 1953.
23. Read, J. C. "The Calculation of Rectifier and Converter Performance Characteristics," *Journal of the IEE,* Vol. 92, Pt. II, 1945, pp. 495–509.
24. Key, Thomas S., and Jih-Sheng Lai. "Comparison of standards and power supply options for limiting harmonic distortion in power systems." *IEEE Transactions on Industry Applications,* Vol. 29, no. 4, July/August 1993.
25. April, G. E., and G. Olivier. "A novel type of 12 pulse converter." *Conference Record,* Industry Applications Society, IEEE-IAS-1982, Annual Meeting 1982, pp. 913–922.

26. PQTN Brief #5, "Harmonic Filter for Personal Computers: Passive, Series Connected, Parallel Resonant," published by the EPAI, Power Electronics Applications Center, October 1992.
27. Enjeti, Prasad, and Ashek Rahman. "A new single phase to three-phase converter with active input current shaping for low cost ac motor drives." *Transactions on Industry Applications,* Vol. 29, no. 4, July/August 1993.
28. Lowenstein, Michael Z. "Improving power factor in the presence of harmonics using low-voltage tuned filters." *IEEE Transactions on Industry Applications,* Vol. 29, no. 3, May/June 1993.
29. Martin, R. "Harmonic distortion and IEC 555-2." *Compliance Engineering,* Fall 1991.
30. Divan, D., B. Banerjee, D. Pileggi, R. Zavadil, and D. Atwood. "Design of an active series/passive parallel harmonic filter for ASD loads at a wastewater treatment plant." *Proceedings of Second International Conference on Power Quality,* sponsored by Electric Power Research Institute, Atlanta, September 28–30, 1992.
31. Rice, David E. "A detailed analysis of six-pulse converter harmonic currents." *IEEE Transactions on Industry Applications,* Vol. 30, no. 2, March/April 1994.
32. Merhej, Saad J., and William H. Nichols. "Harmonic filtering for the offshore industry." *IEEE Transactions on Industry Applications,* Vol. 30, no. 3, May/June 1994.
33. Schlabach, Leland A. "Analysis of discontinuous current in a 12-pulse thyristor dc motor drive." *1989 Conference Record of IEEE,* IAS Annual Meeting, San Diego.
34. Stratford, Ray P. "Harmonic Pollution on Power Systems—A Change in Philosophy." *IEEE Transactions on Industry Applications,* Vol. IA-16, no. 5, September/October 1980.
35. Ortmeyer, T. H., M. Shawky, A. A. Hammam, and J. M. Shaw. "Design of reactive compensation for industrial power rectifiers." *IEEE Transactions on Industry Applications,* Vol. IA-22, no. 3, May/June 1986.
36. Schauder, Colin. "A regenerative two-quadrant converter for DC-link voltage source inverters." *1988 Conference Record of IEEE,* IAS Annual Meeting, Pittsburgh.
37. Bose, B. K. *Modern Power Electronics.* New York: IEEE Press, 1992.
38. Chen, Chingchi, and Deepakraj M. Divan. "Simple Topologies for single phase AC line conditioning." *IEEE Transactions on Industry Applications,* Vol. 30, no. 2, March/April 1994.
39. Funabiki, S., N. Toita, and A. Mechi. "A single phase PWM AC to DC converter with a step up/down voltage and sinusoidal source current." *1991 Conference Record of IEEE,* IAS Annual Meeting, Dearborn, Michigan.

40. Thollot, Pierre A. "Power electronics technology and applications." *1993 Selected Conference Papers,* IEEE Technical Activities Board, IEEE Catalog No. 93CR0100-8.

41. Dorf, Richard C. *The Electrical Engineering Handbook.* Boca Raton, FL: CRC Press, 1993.

Index

F

H

I

K

L

M

U

V

Z

Printed in the United States
3149

9 780780 353947